Build a Clancy

A Step-by-Step Guide to Your First Boat

J. D. Brown and Bob Pickett

INTERNATIONAL MARINE
CAMDEN, MAINE

Published by International Marine

10 9 8 7 6 5 4 3 2 1

Library of Congress Cataloging-in-Publication Data

Brown, Jim, 1933-
Build a Clancy : a step-by-step guide to your first boat / J. D. Brown and Bob Pickett.
p. cm.
Includes index.
ISBN 0-87742-318-0
1. Boatbuilding. 2. Sailboats–Design and construction.
I. Pickett, Bob (Robert C.) II. Title.
VM321.B86 1992
623.8′223–dc20 91-47156
CIP

Questions regarding the content of this book should be addressed to:

International Marine
P.O. Box 220
Camden, Maine 04843

Typeset by A & B Typesetters, Bow, NH
Printed by Arcata Graphics, Fairfield, PA
Design by Janet Patterson
Illustrated by Linda Spicher
Edited by Jim Babb, Sarah Price, and Pamela Salomon
Production by Janet Robbins

Contents

Acknowledgments

The Clancy was developed by Bob and Erica Pickett at Flounder Bay Boat Lumber in Anacortes, Washington. Rich Kolin designed the Clancy. Linda Spicher created the illustrations. Students at Anacortes high school made the exploded drawings. Margaret Backenheimer proofed and indexed the manuscript. The authors photographed the building of Clancy, except where indicated. Jim Babb went to bat for the book. The Center for Wooden Boats in Seattle, Washington, provided a venue for building the early Clancys, while the students of Alternative School #1, Seattle Public School District, and their principal, Ron Snyder, pioneered the concept of Clancy in the classroom.

ONE

Overview:
Building Clancy Piece by Piece

Clancy is a 10-foot sailboat of exquisite form and commanding performance. It meets the requirements of the most demanding professional boatbuilders and sailors. At the same time, Clancy is a sailboat you can put together yourself—even if you're a first-time builder.

No special woodworking or marine engineering skills are required. If you can cook from a recipe or sew from a pattern, you can build Clancy from our directions. We tell you how to put Clancy together piece by piece, step by step. Each step has a list of ingredients, and the building process is fully illustrated with drawings and photographs. The parts in the drawings are numbered, and each number corresponds to a pattern for that part. You cut out the parts using the dimensions on the patterns and secure the parts in position.

Clancy is a forgiving boat to assemble. If you make a mistake, if a piece doesn't quite fit, you simply replace it and try again.

When you're done, you end up with a beautiful sailing vessel that can handle the toughest waters with confidence and grace.

THE BOAT

We are fond of saying that Clancy is the most sailboat you can squeeze out of a 10-foot sheet of plywood. Clancy's light weight (about 85 pounds), large sail area (55 square feet), and sleek hull (designed to plane when the wind is right and the load is light) deliver exciting sailing speeds. At the same time, Clancy is trim and petite enough to cartop.

Figure 1-1. The building process consists of these six basic stages.

Stage 1: Build a jig to support the first pieces of the boat.

Stage 2: Assemble the skeleton on the jig. The skeleton consists of the stem, transom, bulkheads, and daggerboard case.

Stage 3: Join the bottom and the two sides to the skeleton on the jig.

Stage 4: After removing Clancy from the jig, complete the interior and finish by sanding, scraping, and fiberglassing the seams.

Stage 5: Add the deck and coat Clancy with varnish or paint.

Stage 6: Rig Clancy with rudder, daggerboard, mast, boom, sail, hardware, and rigging. Then sail away . . . (Photo by Matt Brown)

Quick on the water, Clancy is nevertheless quite safe because it contains large flotation compartments forward and aft. While other sailboats of this size tend to turn turtle or swamp, Clancy remains a stubbornly dry vessel, ideal for anyone learning to sail or for small-boat sailors who hate to get dunked. You'll be able to sail your Clancy without a wetsuit in areas where the water is cold. These ample flotation chambers also double as lockers for dry storage.

Clancy is small, quick, strong, and beautiful, with plenty of room in the cockpit for two (a pair of adults, a pair of kids, or a parent and a child).

Clancy is such a high-performance sailboat that it belongs to its own official class. You can sail it in Clancy regattas for pleasure or for competition. If you build your own Clancy from our instructions, you can become a charter member in the National Organization. (Just complete the registration form at the end of this book.)

YOU, US, AND A DOG NAMED CLANCY

We developed Clancy so you could build a superb wooden sailboat the first time out, no matter who you are or what your skills might be. In so doing, we're taking a solid swipe at the black magic of boatbuilding. Clancy is for people who don't consider themselves boatbuilders at all. The ability to build a sailboat isn't a matter of sex or age. Clancy is a project for any man, woman, or child who ever dreamed of building their own sailboat, but veteran do-it-yourselfers, experienced carpenters, and Old Salts, too, will enjoy the challenge.

One of this book's authors is a veteran boatbuilder, but the other had never built a boat until he was roped into this project. The result is a builder's guide quite literally written from the perspective of a first-time boatbuilder. He has made doubly sure that everything is spelled out, down to the last screw hole and third coat of varnish. Frankly, if our writer could build Clancy, then so can you.

Clancy has entered the classroom, too. At the Alternative School #1, which draws kindergartners through eighth graders from the Seattle Public School System, seven Clancys have been scheduled to be built as class projects. Principal Ron Snyder has seen his young students so taken with this assignment that he helped them found one of the first all-student public yacht clubs in the United States (with 13-year-old La Vonne Beaver serving as the first commodore).

In short, Clancy is ideal as a sail trainer, a classroom project, a home project for parent and child, or an introduction to wooden boatbuilding for any first-time builder.

Clancy is easy enough to build from our recipe, but that's not to say this is a sailboat you can rip and hammer together overnight. No great boat is assembled without an investment in time and an expenditure of patience. Under the guidance of an experienced boatbuilder, Clancy can be completely assembled by an amateur class in as little as two weekends. We have done it this way at the Center for Wooden Boats in Seattle. When you do it yourself for the first time, with this book as your expert boatbuilding guide, it will take longer, of course, but it's by no means difficult for anyone willing to take the plunge. In the end, you'll have learned plenty about the art of boatbuilding; you'll also have a vessel you can enjoy for years afterward.

In technical terms, Clancy is built using the stitch-and-glue method. In plain English, this means you stitch the boat together on a frame (the jig) using glue, screws, and fiberglass tape on the seams. Clancy has just 29 basic pieces, each cut from wood using the dimensions we give you on the numbered patterns. Each piece has its own number, pattern, and set of directions. Step by step, Clancy takes shape.

When you finish, you'll find that Clancy's high-tech materials and construction techniques will keep her looking beautiful for years with a minimum of upkeep. Fine wooden sailboats need not be a headache to maintain, provided they are built with the proper supplies and methods, and Clancy is among the best of the small sailboats on the water today.

So, if you've built boats before, you'll find Clancy a project worthy of your expertise; if this is your first time out, Clancy will make admirers think you're an expert builder.

Should you need advice about materials or building

techniques, just call Bob or Erica Pickett at Flounder Bay Boat Lumber on the Clancy Hotline (1-206-293-2369). We developed Clancy from its inception, with the philosophy that there's simply no better sailboat afloat than the one you build yourself, and we'll be glad to answer your questions as you begin building.

This is our second boat book. We started out in 1985 with a 14-foot rowboat known far and wide as the Cosine Wherry. More than three thousand of you have purchased our first book so far, and hundreds have built their own Cosine Wherries from our directions. Clancy is smaller, easier to build, just as pretty as the Cosine Wherry–and best of all, it sails. We developed Clancy in response to the need for a safe, small, lightweight, high-performance sailboat of distinction that was fully within the reach of any beginning builder.

Clancy's designer, Richard S. Kolin, based in Puget Sound, has been a master at creating sleek and economical small sailboats for more than 20 years. Rich bought his first sailboat, a 7-foot skiff, at age nine. His early years learning to sail inspired Clancy, which we consider his masterpiece.

As for Clancy–well, Clancy was our designer's loyal dockside dog, an exuberant canine and dedicated sailor, always underfoot whenever Rich was building or launching his boats around Puget Sound. We thought about giving this high-performance sailboat a suitably high-profile name, but Clancy was the name that stuck from the beginning. That's why we emblazoned the Clancy Class sail with its distinctive insignia–the dog bone. No doubt it would have appealed to the real Clancy. But the Clancy you build, whether under sail or on the dock, is no dog, as you'll quickly discover.

Whatever your past doubts about being able to build a fine wooden sailboat, we now invite you to give this recipe a try. It's complete, simple, and foolproof–and in the end, you'll have something you can enjoy and admire for a lifetime. If you can cut out a piece of wood and join it to other pieces using glue and screws, Clancy is yours for the building. All it takes is some ambition and patience, plus the right tools, ingredients, and recipe.

TOOLS

The tools you need to put Clancy together are common and easy to use. They include:

Bar clamps (2)

Barrier cream

Clamps (4 spring clamps and 2 C-clamps, 3½-inch size)

Countersink and stop collar for drill

Drill (reversible)

Drill bits (1 standard set, $\frac{1}{16}$- to ¼-inch size)

Dust mask

Hammer

Handsaw

Hot-melt glue gun

Keyhole saw

Paint brushes

Paper cups, towels

Pencil

Plane (block style)

Plastic gloves

Putty knife

Saber saw (portable jigsaw)

Safety goggles

Sander (speedblock)

Sawhorses (2)

Scissors

Scraper

Screwdrivers

Shoe file

Skilsaw

Squeegee

Square (large size)

Tape measure

Twine

Wooden dowel

Wood rasp

INGREDIENTS

Basic ingredients for cooking up a Clancy are available from local lumber retailers, marine hardware outlets, and paint stores. The list includes the following:

Lumber

- 2 × 4 × 10-foot straight fir, spruce, hemlock, or pine (5 boards) select or #2 grade–for the jig
- 2 × 2 × 12-foot clear fir (2 boards)–for jig, log, blocking

- 1 × 4 × 12-foot fir or mahogany (6 boards)—for keelson, keel, guardrails, all doublers
- 1 × 2 × 12-foot clear fir (1 board)—for cutwater, daggerboard-case strips
- 1/2-inch ACX 5-ply plywood or particle board (1 sheet, 4 × 4 feet)—for center mold, crutch, gussets
- 1/2-inch 5-ply mahogany or fir marine plywood (1 sheet, 4 × 4 feet)—for transom, deck beam, kingplanks, daggerboard case (see "Fir Versus Mahogany" on page 10)
- 1/4-inch 5-ply mahogany or fir marine plywood (3 sheets, 4 × 10 feet)—for sides, bottom, bulkheads, partial bulkheads, carlins, deck, deck trim, rudder
- 3/4-inch 7-ply mahogany marine plywood (1 sheet, 2 × 4 feet)—for daggerboard, mast step
- 2 × 3 × 4-foot clear fir, mahogany, or oak (1 board)—for daggerboard handle, tiller
- 2 × 6 × 3-foot clear fir (1 board)—for transom knee, stem

Note: The dimensions we give for lumber (except for plywood) are the standard nominal lumberyard dimensions, but the actual lumber you buy will be milled so that the pieces are smaller. (A 2 × 4, for example, actually measures closer to 1 1/2 inches by 3 1/2 inches.)

Other Materials

- Acetone
- Fiberglass cloth, 60 inches wide, 6-oz. wt. (12-yard length)—*only if you use fir instead of mahogany marine plywood*
- Fiberglass tape, 3 inches wide, 6-oz. wt. (50-yard roll)
- Marine varnish, marine paint
- Plastic flanges, circular, 8 1/2-inch-diameter (3 sets)—to cover access ports in bulkheads
- PVC pipe, 2-inch diameter, schedule 40 (2 feet long)—for mast tube

- Sandpaper (assorted grits, including 100-grit 3M Trimite for sanding epoxy and fiberglass cloth)
- 1/8-inch marine aluminum (1 sheet, 2 × 2 feet)–for rudder plate
- 1/16-inch piece of stiff but flexible plastic or an existing squeegee–for filleting tool

FIR VERSUS MAHOGANY

You must use a high-quality marine plywood to build your Clancy, but should you choose 3-ply fir or 5-ply mahogany? In a nutshell, the differences between fir and mahogany are (1) strength and weight, and (2) time and money.

Fir marine plywood is weaker than mahogany. This means that you must strengthen a boat built of fir by adding more fiberglass sheathing. Fir also *checks*–that is, develops unsightly marks and breaks as it shrinks–unless protected by a coating of fiberglass. Any fir marine ply in the boat must be fully fiberglassed. Mahogany, on the other hand, requires fiberglass taping only at the seams. Thus, mahogany has two inherent advantages over fir: It is a stronger material, and it resists checking.

Once you coat the fir with fiberglass, it is virtually as strong as mahogany. The extra fiberglass does add a few pounds to Clancy, but the main drawback to extra fiberglass is the time it requires to apply. If you use fir instead of mahogany, count on spending 10 to 15 percent more time building your boat.

The bottom line is cost. Fir marine ply is considerably less expensive than mahogany marine ply. You might save $150 or more by going with fir. This price gap narrows, however, when you factor in the cost of extra fiberglass and epoxy resin. Ultimately, you will save $100 or so by going with fir plywood.

If you can afford it, go with mahogany; if not, trust in the fir. Either material will look good and hold up for many years. Both materials are reliable and have been used for fine wooden sailboats by generations of master builders.

In this book, we give you a recipe for building Clancy with either of the two marine plywoods, inserting a few extra steps at times for those of you who are using fir.

Hardware and Adhesives

To hold all the above materials in place, you will need the common fasteners and glues in quantities that vary from one builder to the next. The key ingredients in this category are:

- Bolts, nuts, screws, washers, and fittings for rudder as specified in Chapter 6
- Bronze screws or stainless steel screws, flathead (about 100 each in sizes 1-inch × #6, 1-inch × #8, and 1¼-inch × #8, and a dozen in sizes ⅝-inch × #6, ¾-inch × #6, ¾-inch × #8, 2½-inch × #10, and 2½-inch × #12)
- Carpenter's glue
- Epoxy resin and hardener (about 1.5 gallons)
- Wood flour
- Finishing nails (1-inch and 3-inch lengths)
- Glue sticks (4 pieces)–for hot-melt glue gun
- Marine adhesive sealant (urethane or polysulfide elastomeric–*no silicones*–brown or mahogany color)
- Rigging (cleats, eye straps, lines, et al.) for sail, boom, and mast as specified in Chapter 7
- Sheetrock screws (about 50 in 1-inch, 1½-inch, 2-inch, and 2½-inch lengths)

Estimated Costs

As of 1992, you could expect these ingredients to cost about as follows (depending on choice of materials):

	FIR	MAHOGANY
Lumber	$300	$450
Other Materials	$200	$180
Hardware, adhesives	$150	$120
	$650	$750

In addition, the sail, mast, boom, and rigging, when purchased from a retail dealer, could cost another $750 or so. The bottom line for your Clancy should be under $1,500–a rock-bottom price for a fine wooden sailboat.

WORKING SAFELY

Protect yourself and the environment. Work cleanly and safely. Always use safety goggles when working on Clancy, and be sure to read the labels on epoxies, paints, adhesives, and other chemicals you use in boatbuilding. Consult retailers for directions and safety instructions. Above all, always maintain a protective screen between yourself and epoxy resins, sealants, fiberglass, paints, and varnishes either by applying barrier cream to your hands and arms or wearing plastic gloves. Don a mask when sanding or scraping. A fume mask is recommended if you are particularly sensitive to fumes, dust particles, or the chemical sensitizers used in the hardening agent you mix with epoxy. Of the chemicals, acetone is the most dangerous because it is very volatile. Be sure to wear a charcoal-filter mask when working with sealants, and protect yourself with either plastic gloves or barrier cream. The same goes for fiberglass: Use a dust mask or fume mask, and protect your skin with cream or plastic gloves.

While washing up, avoid solvents and acetone. Rely on waterless handcleaners. Warm water and soap are best. For disposal of chemicals (epoxy, paint, etc.), check with your local environmental authorities. If you mix hardener with epoxy and let it stand, the substance solidifies; paint solidifies if you leave the can open. In both cases, the solidified mass can be dumped – but only if your local environmental regulations allow it.

TWO

The Jig:
Something to Build On

The jig is the frame that holds Clancy together while you assemble some of the most basic pieces, such as the bottom and the sides. The jig consists of four main pieces:

1) a frame aptly called the *ladder* (Part 1);

2) a wooden strip, in three sections, fastened to the ladder rungs, called the *log* (Part 2);

3) a slotted plywood form, fastened to the log, called the *crutch* (Part 3);

4) another plywood form, inserted in the crutch and fastened to the ladder, called the *center mold* (Part 4). There are also some minor support pieces called *gussets* and *blocking*.

In Figure 2-1, an exploded view of Clancy's jig, you can see how the pieces fit together.

Think of the jig as the temporary skeleton for the boat. Once finished, it looks like the remains of a shark. Your goal is to build a strong jig. It must not crumble under the pressure of the pieces you bend around it.

The jig is simple but time consuming to build. None of the pieces of the jig are a permanent part of Clancy, but all are necessary to begin the recipe.

Building the jig is sometimes called the least-rewarding phase of boatbuilding. Fortunately, it must be done first, while our enthusiasm is high. The subsequent stages are more rewarding because Clancy begins to look more and

TOOLBOX

Bar clamps (2)
Clamps (spring clamps, C-clamps)
Drill (reversible)
Drill bits
Hammer
Pencil
Saber saw
Sawhorses (2)
Square
Tape measure

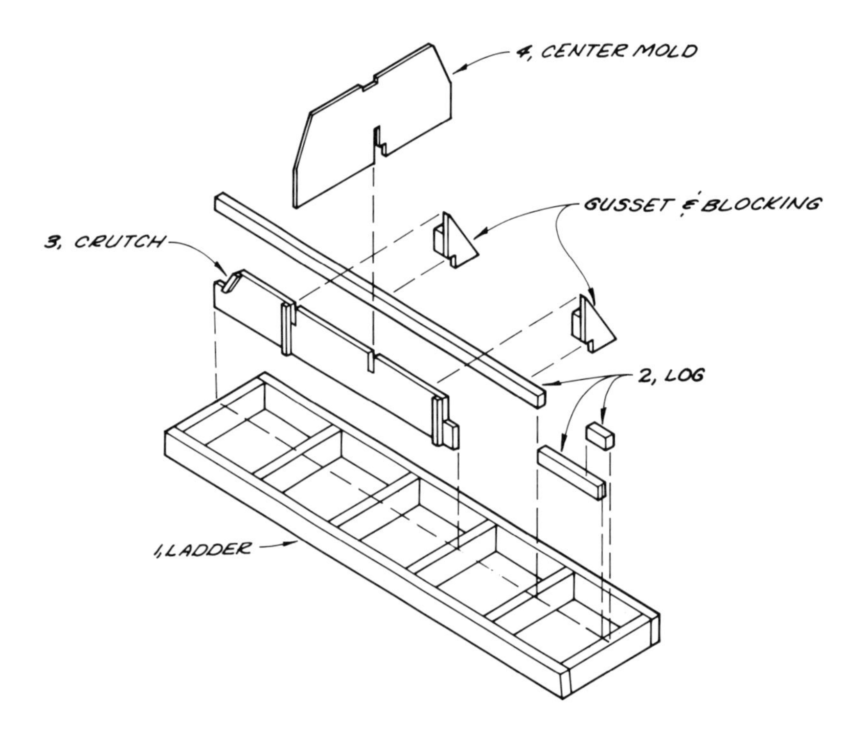

Figure 2-1. Exploded view of the jig.

more like the boat you'll soon be sailing. For now, you're confronted with the thankless groundwork—thankless until you lend the jig to a friend.

Orientation: *Forward* is always toward the front of the boat; *aft* is always toward the back of the boat. Since Clancy builders tend to be forward-looking people, we usually orient our instructions in that direction.

Tip: Take it one piece at a time.

PART 1: LADDER

The ladder is the base of the jig (Figure 2-2). The sides of the ladder are exactly 10 feet long. Each of the six rungs is 3 feet wide. The gussets are triangular pieces of plywood added at the intersections for support.

Ingredients

- 2 × 4 straight lumber (two 10-foot sides, six 3-foot rungs)
- 1/2-inch ACX 5-ply plywood or particle board (gussets)

- Nails (3-inch finish nails)
- Carpenter's glue

Directions

1) Cut out two 10-foot sides and six 3-foot rungs.

2) Lay the side pieces out on sawhorses.

3) Spread glue where the end rungs (#1 and #6) touch the sides. Then drive two nails through each side of the ladder into the ends of the #1 and #6 rungs.

4) Clamp rungs #1 and #6 between the ends of the ladder sides with bar clamps (Figure 2-3).

5) Label rung #1 *aft* with a pencil and rung #6 *fore*. Knowing which end is which is crucial, even in boatbuilding.

6) Moving forward, glue and nail into place the rest of the rungs at the exact positions shown in Figure 2-4. Check for squareness as you go.

7) You can cut gussets from new plywood, particle board, or scraps. Twelve-inch squares work well: just cut across diagonally and you have two triangular gussets. Glue and nail a gusset into position wherever a rung meets a side (Figure 2-5).

8) Flip the ladder over so that the gussets face earthward.

9) Last but not least, draw a centerline lengthwise across the rungs of the finished ladder.

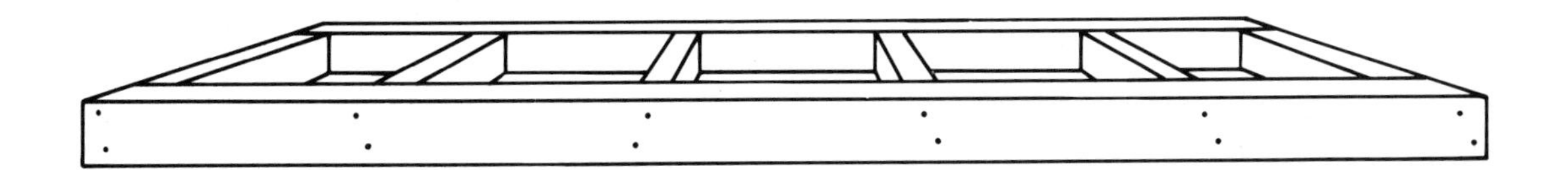

Figure 2-2. The ladder frame is the foundation of the jig.

Figure 2-3. With rungs #1 and #6 secured at each end, the ladder is clamped square.

Tips: Check as you go with the big square to be sure everything stays lined up. All that hammering can throw the ladder out of square.

Assembling the ladder on two sawhorses works best and creates a good working height.

Use bar clamps to hold the frame together as you hammer.

Figure 2-4. Proper location of the rungs, gussets, and centerline on the bottom of the ladder.

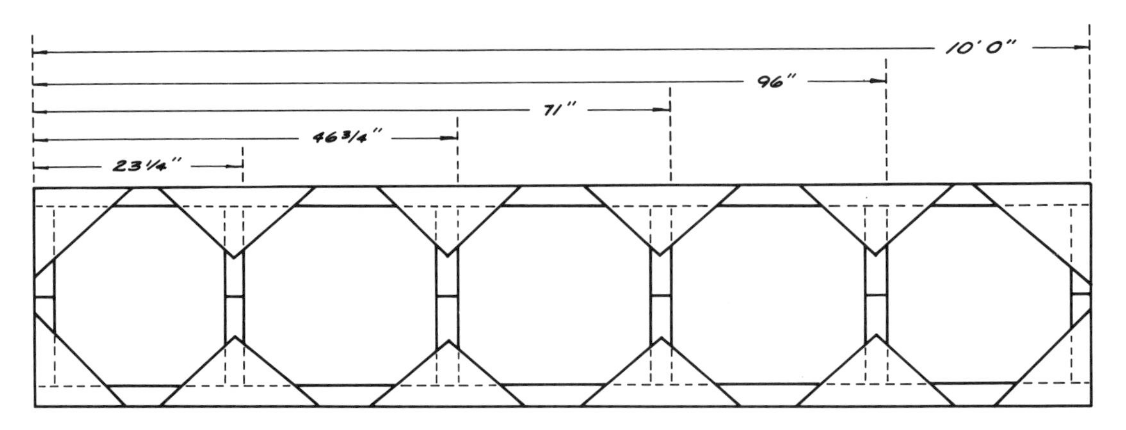

VIEW OF THE BOTTOM SIDE OF LADDER FRAME
CENTERLINE SHOWN
PART #1

Figure 2-5. With gussets attached to the rungs, the ladder is ready to be flipped over.

PART 2: LOG

The log runs the length of the ladder and provides a means to attach other pieces to the jig. The log consists of one long section (#1) and two short sections (#2, #3) placed between the last set of rungs. Figure 2-6 shows the proper layout of these sections of the log when attached to the ladder.

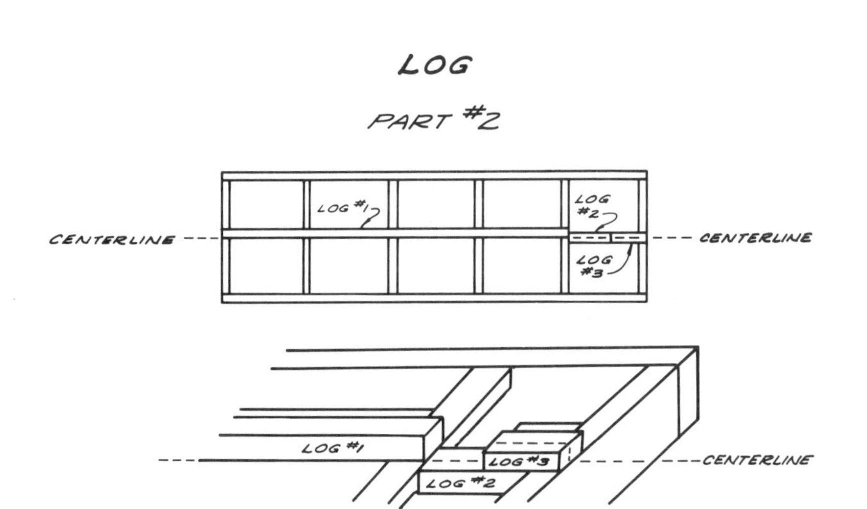

Figure 2-6. The log consists of three sections, positioned along the centerline of the ladder.

Ingredients

- 2 × 2 clear fir (12 feet long)
- Nails ($2^1/_2$- to 3-inch finish nails)
- Carpenter's glue

Directions

1) Cut out log #1 (the 8-foot section in Figure 2-6) and position it on the left side of the centerline (as you look *forward*). The log should just touch the centerline along its right side (looking *forward*).

2) Glue and nail log #1 in position on the ladder, beginning *aft*. Tip: Drilling some pilot holes in the log, using a slightly smaller bit than the diameter of the nails, makes for easier and lighter hammering.

3) To fill in the final gap at the *forward* end, measure the space (about $10^1/_2$ inches) between the rungs and cut out log #2 to fill it. Draw a centerline the long way on this piece (log #2) and nail it into position so that its centerline matches that on the ladder. Log #2 should fit snugly between, not on top of, the final two rungs. Log #2 is recessed below and offset from log #1. It acts as a brace for log #3.

4) Cut log #3 from the same material, 12 inches long. Draw a centerline. Glue and nail it on top of log #2 and the final rung, centered as shown in Figure 2-6.

PART 3: CRUTCH

The crutch is a part of the jig (not part of Clancy). It is attached to the log and supports other pieces. Three strips of 2 × 2, called blocking, are mounted on the crutch.

Ingredients

- $^1/_2$-inch plywood or particle board (crutch)
- 2 × 2 fir, short lengths (blocking)
- Sheetrock screws (1 inch to $1^1/_2$ inches long)
- Carpenter's glue

Directions

1) Since the crutch's sole reason for existence is to position other parts of the jig, it must be cut exactly to the dimensions shown on the crutch pattern (Figure 2-7). Copy the pattern onto wood and cut the outline with a saber saw.

2) Check that the notches are wide enough by placing a scrap of 1/4-inch or 1/2-inch plywood in them.

3) Cut out and glue three blocking strips to the crutch as shown in Figure 2-8 (one flush with the slanting *aft* end, one along the *aft* edge of the first slot, and one along the *aft* edge of the third slot of the crutch). Screw through the plywood crutch into the blocking.

4) The crutch is attached to the log on the right side as you look forward (Figure 2-9). Begin flush with the *aft* end of the log. Glue the crutch to the log and clamp into position. Then drive screws, set every 6 to 8 inches, through the plywood crutch into the log.

Figure 2-7. The crutch is attached to the log.

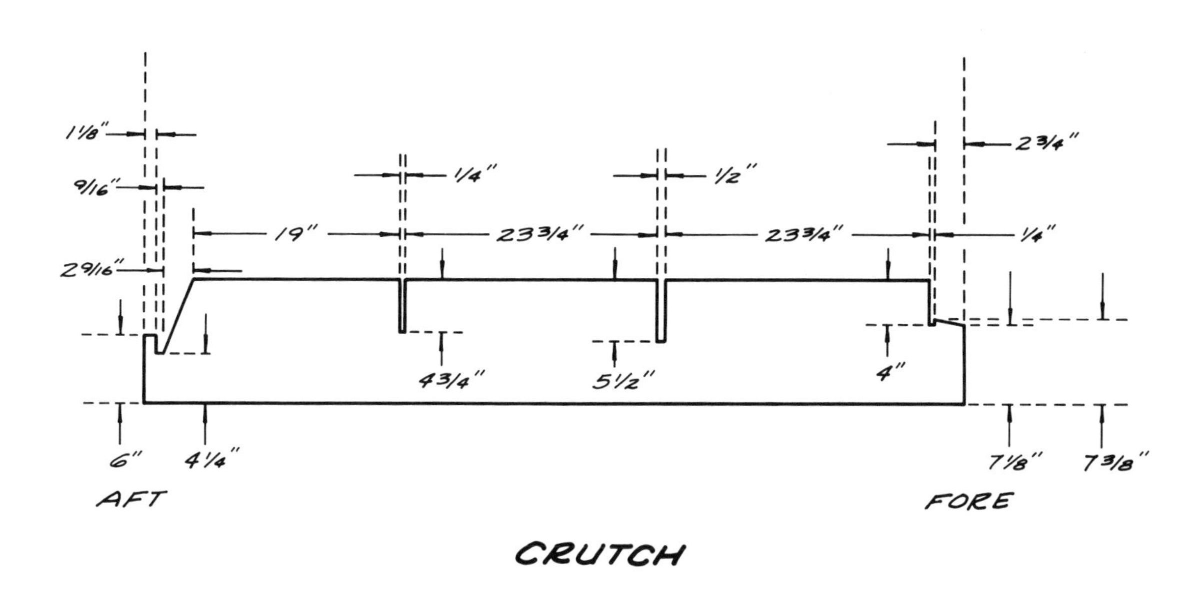

Figure 2-8. Attach three blocking strips to the crutch (looking forward).

Figure 2-9. The crutch is glued and screwed to the log on the starboard side (right side looking forward).

GUSSETS WITH BLOCKING FOR THE CRUTCH

Gussets are triangular plywood supports which hold the crutch securely in place and square to the jig. Blocking is a length of 2 × 2 lumber to which other pieces are fastened. Two gussets with blocking are attached to the jig.

Ingredients

- 1/2-inch plywood (gussets); scrap pieces are okay
- 2 × 2 fir (blocking attached to gussets)
- Sheetrock screws (2 inch to 2 1/2 inches long)
- Carpenter's glue

Directions

1) Cut out two triangular gussets from plywood. Each gusset should be about 10 inches tall and 6 inches wide.

2) Notch each gusset where the two legs of the triangle meet so that the gusset will fit in place over the log. The notch is roughly 1 1/2 inches by 1 1/2 inches square.

3) Cut out two lengths of blocking, each short enough to fit above the gusset notch. Glue and screw one blocking strip into each gusset. Screw through the plywood gusset into the blocking.

4) Looking forward from the back of the ladder, position one gusset against the *aft* (back) edge of the second rung. This gusset is on the left (port) side of the crutch; its blocking strip is on the *aft* (back) side of the gusset, touching the plywood crutch. Clamp the gusset piece into place.

5) Position a second gusset on the forward side of the fourth rung. Clamp it. Blocking is *aft* of the gusset. Figure 2-10 shows you the overall layout.

6) With everything in position, glue the two gusset blocking strips to the plywood crutch. Screw through the crutch into each gusset blocking strip at two points.

7) Using the big square, check that the ladder is square with the crutch.

8) When everything is squared up, finish securing the two gussets by screwing through them into the rungs.

Tip: Using a reversible drill with a screwdriver bit can speed up the setting of screws and save on hand strain. If you haven't used a drill to set screws before, it may take some

Figure 2-10. Fasten the gussets to the crutch on the port side (left side, looking forward). Then square the crutch to rungs #2 and #4 and fasten the gussets there to hold everything square.

practice. Building the jig is a good place to practice. Begin by drilling shallow pilot holes (slightly smaller than the diameter of the Sheetrock screw). Then put the screwdriver bit on your drill and slowly drive the screws in. Remember to lean down hard on the drill and turn the shaft very very slowly. Don't race the drill motor. By reversing the drill, you can also back screws out.

PART 4: CENTER MOLD

The last piece of the jig–the center mold–is a crosspiece that supports the keelson and the sides of the boat. It slips into the middle slot of the crutch.

Ingredients

- 1/2-inch plywood or particle board (center mold)
- 2 × 2 strips, 18 inches long (for two blockings)
- Sheetrock screws (1 inch to 1 1/2 inches long)
- Carpenter's glue

Directions

1) Transfer the center mold pattern (Figure 2-11) onto plywood and cut out with a saber saw.

2) Be sure the center mold slips into the correct slot on the crutch (over the third rung). Draw a vertical centerline on the center mold.

3) Cut two 18-inch strips of blocking to fit on either side of the center mold's slot (where it fits over the log). Attach the blocking strips on the *aft* side of the center mold with glue and 1-inch Sheetrock screws. Drive screws through the center mold into the blocking.

4) Slip the center mold into its slot. Line up the centerlines on the center mold and the ladder so that they meet. Lift slightly and apply glue along the third rung. Lower the center mold back into position. Check that the center mold is square with the crutch. Clamp the center mold in place.

Figure 2-11. The last piece of the jig, the center mold supports Clancy's keelson and sides.

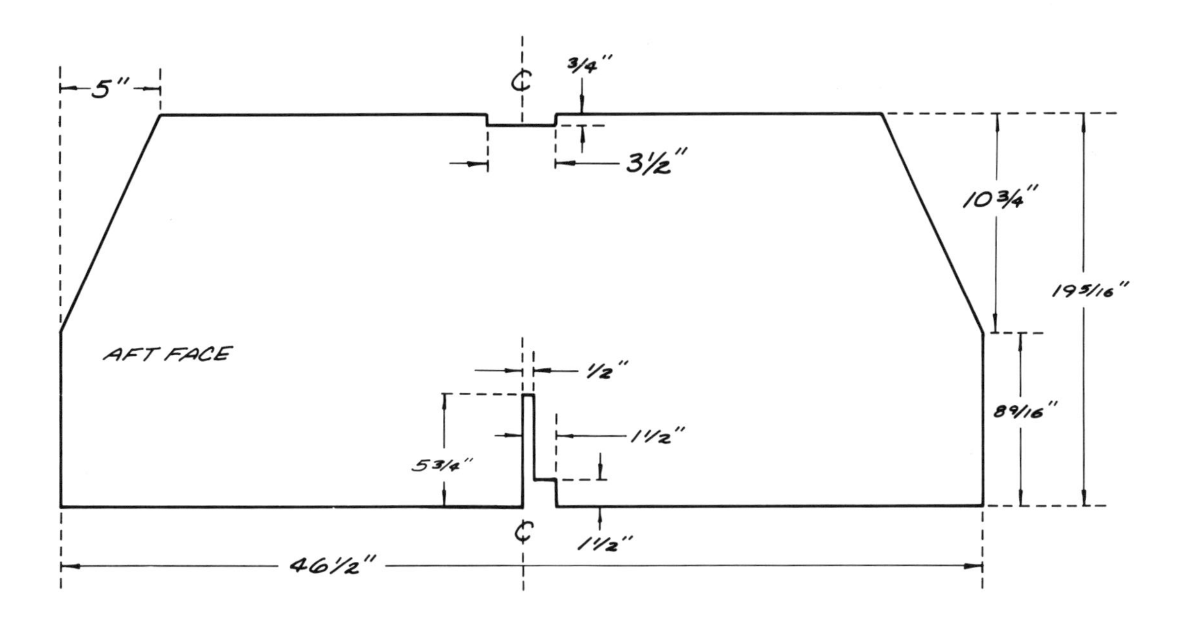

CENTER MOLD

PART #4

Figure 2-12. The center mold is the last part of the jig.

5) Drive a screw every 6 inches or so through the center mold blocking strips into the third rung of the ladder. Your center mold should fit in the crutch as shown in Figure 2-12. Congratulations! The jig is up! Now it's time to add the first permanent pieces to Clancy.

THREE

The Skeleton:
From Stem to Stern

Now that the jig's in place, you can add the first pieces to Clancy proper–the boat's backbone (the keelson and stem) and its ribs (the bulkheads and transom).

Clancy's skeleton consists of a narrow stem up front and a trim transom aft. These pieces are connected by a long thin keelson. Attached along the way to the keelson are a daggerboard case and two bulkheads, one forward, one aft.

You are assembling Clancy upside down at this point. The transom rests in a ledge on the very back of the jig's crutch. The two bulkheads slide into the slots fore and aft of the center mold. The keelson is connected to the transom at the back and to the stem at the front. The stem is connected, temporarily, to the small log section on the front of the ladder. The daggerboard case is connected to the keelson and the forward bulkhead.

All these skeletal connections are a bit baffling in words, so we have provided an exploded view of the whole assembly in Figure 3-1, along with some proverbial thousand-word pictures (Figures 3-2 and 3-3).

TOOLBOX

- Clamps (spring clamps, C-clamps)
- Drill (reversible) with countersink and stop collar
- Drill bits
- Keyhole saw
- Paint brush
- Pencil
- Saber saw
- Square
- Tape measure
- Twine

PART 5: FORWARD BULKHEAD
PART 6: AFT BULKHEAD

The two bulkheads slip into the slots on the crutch, one forward and one aft of the center mold. The bulkheads are permanent parts of Clancy and provide cross support as well as compartment walls for flotation and storage under the deck.

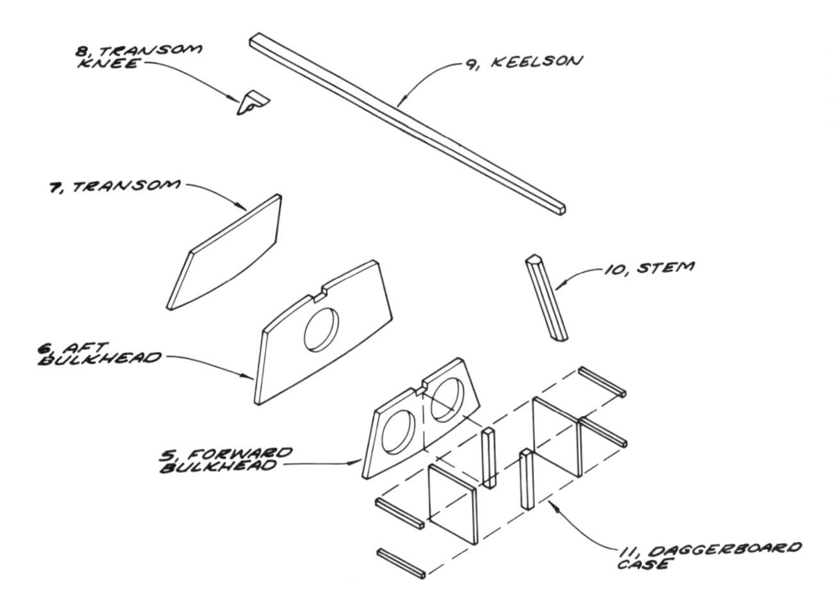

Figure 3-1. Exploded view of the skeleton.

Figure 3-2. The transom and keelson, two parts of the backbone in place on the jig (looking forward).

Figure 3-3. The stem, daggerboard case, and keelson (three parts of the backbone) in place on the jig (looking aft).

Ingredients

- $1/4$-inch 5-ply marine plywood, fir or mahogany
- 3 plastic flanges ($8 1/2$-inch diameter, circular)
- Sheetrock screws (1 inch long)

Directions

1) Using the patterns (Figures 3-4 and 3-5), carefully cut out both bulkheads. Be sure to cut out the flange holes (called access ports)–one in the *aft* bulkhead and two in the *forward* bulkhead. (The first time the author built a Clancy, he forgot to cut the access ports, as perhaps you can see from our pictures; the openings were much harder to cut once these pieces were inside the boat.) To cut a circular opening, you need a good saber saw, a good keyhole saw, or a good friend with the right tool. Check that the flanges fit the holes you cut.

2) Draw vertical centerlines on both bulkheads.

3) Position the bulkheads (the *forward* one goes for-

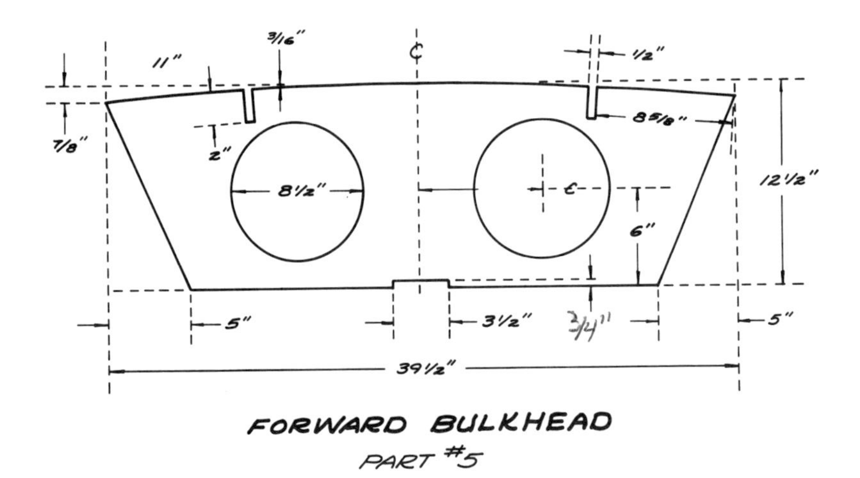

Figure 3-4. The forward bulkhead is placed on the jig, but later, when Clancy is removed from the jig, the bulkhead becomes a permanent part of the boat.

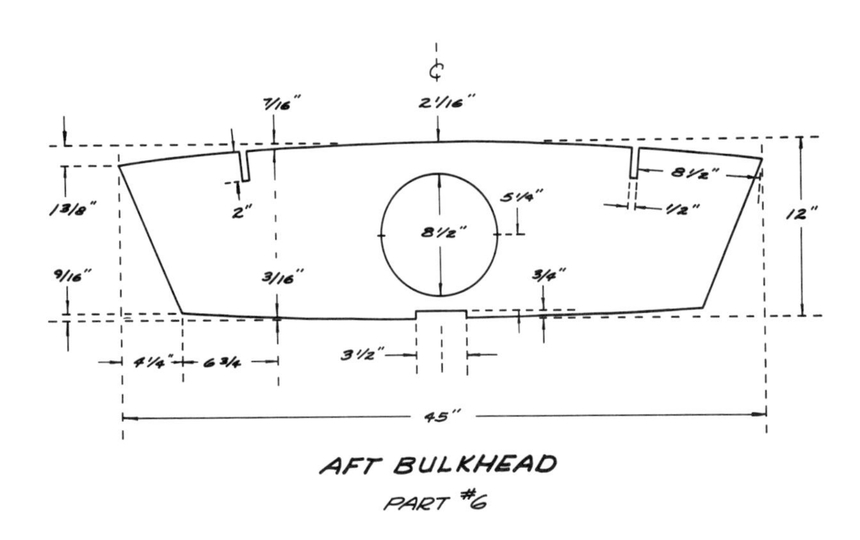

Figure 3-5. The aft bulkhead is placed on the jig, but later, when Clancy is removed from the jig, the bulkhead becomes a permanent part of the boat.

ward, of course). Line up the centerlines on the bulkheads with the slots on the crutch. Figure 3-6 shows the correct positions.

4) Screw the two pieces into place temporarily, driving through the bulkheads into the blocking strips on the crutch. That's it for now.

Figure 3-6. Position the forward and aft bulkheads on the jig.

PART 7: TRANSOM
PART 8: TRANSOM KNEE

Clancy's rear end, the transom, is a permanent and prominent fixture, and should be constructed of fine marine plywood. A small brace, called the transom knee, is also required.

Ingredients

- 1/2-inch 5-ply mahogany or fir marine plywood (transom)
- 2 × 6 × 1-foot clear fir (transom knee)
- Bronze screws (1-inch × #8)
- Carpenter's glue

Directions

1) Cut out the transom and transom knee with a saber saw, transferring the dimensions from the patterns (Figures 3-7 and 3-8).

2) Draw a vertical centerline on both faces of the transom and on the flat edges of the transom knee.

3) Position the transom on its ledge on the *aft* end of the crutch (Figure 3-9). Use the transom centerline to achieve correct position with the edge of the crutch. Drive two bronze

screws through the transom into the blocking strip on the crutch.

4) Position the knee on the forward face of the transom, 3/4 inch below the top, so that the centerlines meet. Use a scrap of stock 3/4-inch fir to be sure that the knee is the proper distance below the top of the transom.

5) Glue the knee to the transom when it is centered. Then drive two bronze screws through the transom into the knee.

6) Bevel the top edge of the transom with a plane so that it is at the same angle as the top of the knee. (The top edge of the transom should be at the same horizontal angle as the top of the knee.) Beveling the top edge of the slanted transom creates a horizontal surface that will eventually accept the bottom.

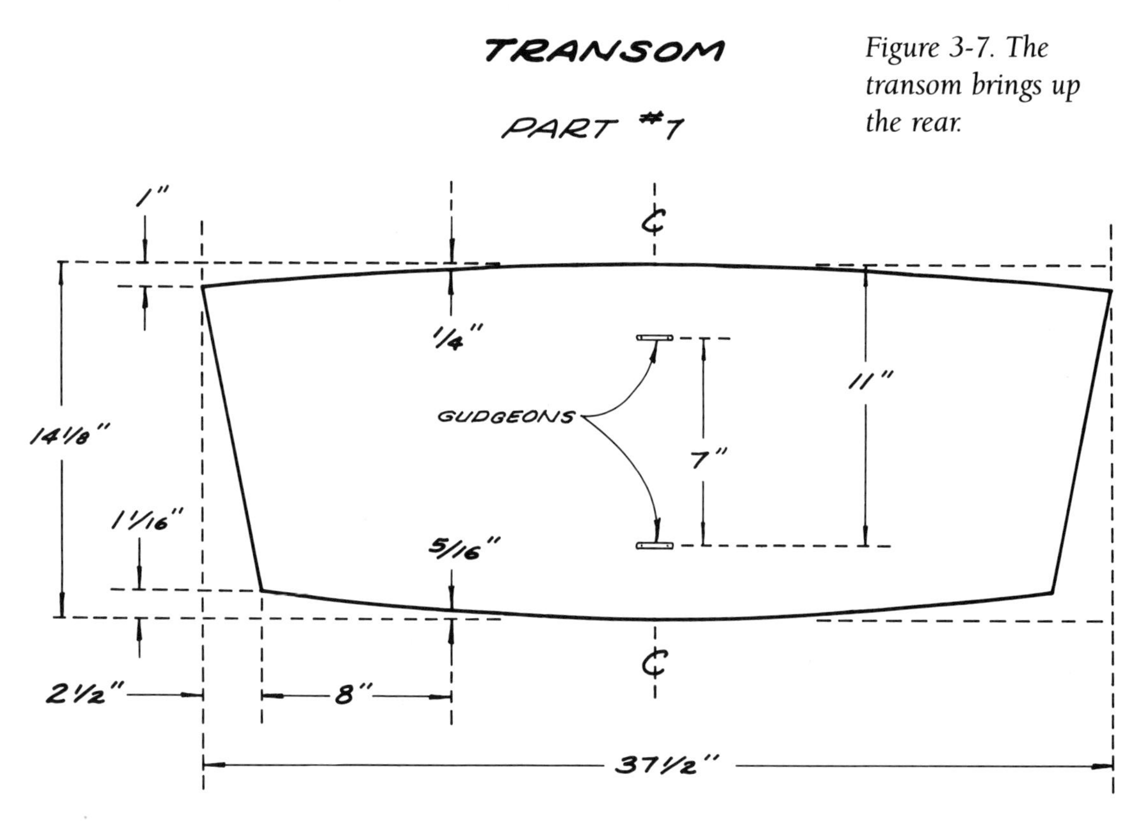

Figure 3-7. The transom brings up the rear.

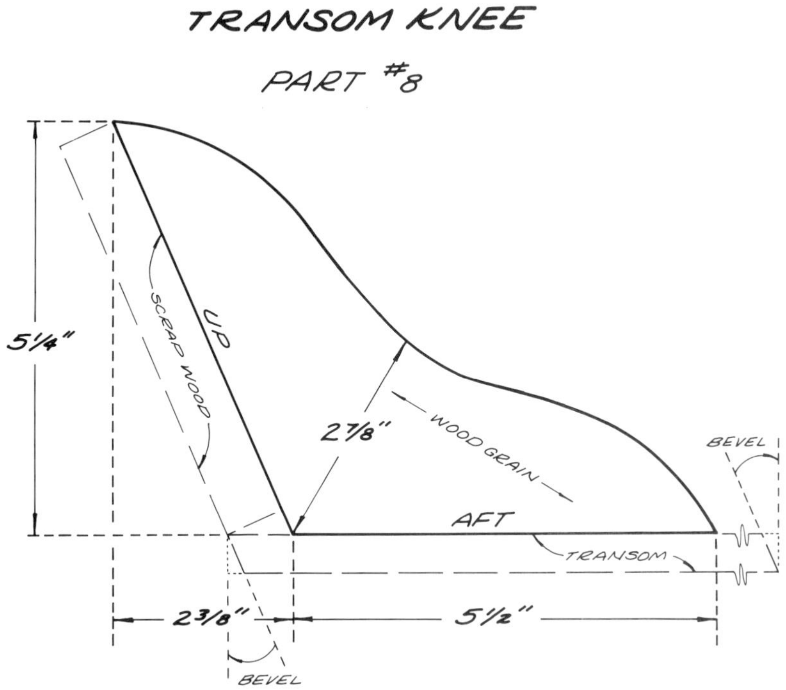

TRANSOM KNEE

PART #8

UP

SCRAP WOOD

WOOD GRAIN

AFT

TRANSOM

BEVEL

5¼"

2⅞"

2⅜"

5½"

BEVEL

Figure 3-8. The transom knee fits on the beveled transom as shown. Note the "aft" and "up" designations.

Figure 3-9. Place the transom in the slot of the crutch on the aft end of the jig (looking forward).

PART 9: KEELSON
PART 10: STEM

The keelson and stem are two separate pieces, but are positioned together on the jig and can be built as a single extension of Clancy's backbone.

Ingredients

- 1 × 4 fir or mahogany, $105\frac{3}{4}$ inches long (keelson)
- 2 × 4 clear fir, $17\frac{1}{8}$ inches long (stem)
- Bronze screws ($1\frac{1}{2}$-inch × #10)
- Carpenter's glue
- Twine

Directions

1) Cut out the keelson and the stem, using the above ingredients and the dimensions on the patterns (Figures 3-10 and 3-11). The stem must be shaped as in the pattern (see the cross section in Figure 3-11).

2) Join the tapered end of the keelson to the bigger end of the stem with plenty of glue and one bronze screw. Place the screw as far back from the keelson tip as possible.

Figure 3-10. The keelson is the spine of Clancy's skeleton.

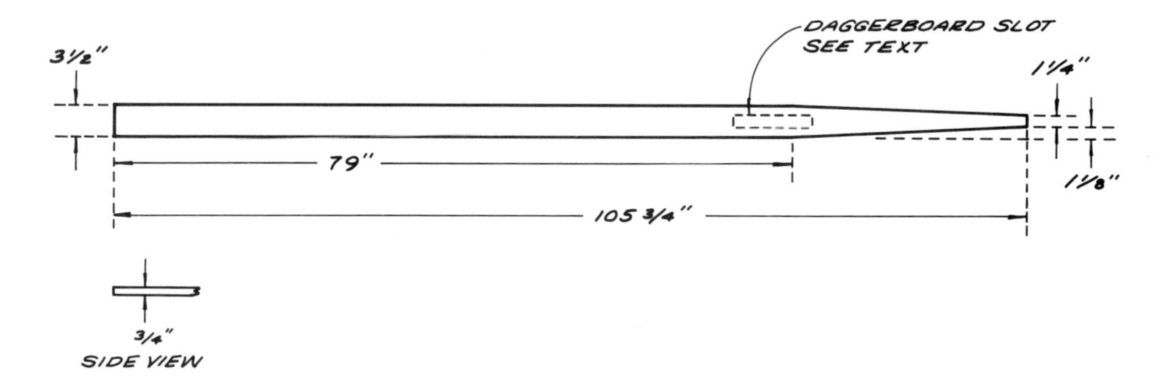

3) Drape the keelson and stem over the jig, line up the aft end of the keelson so it fits completely on the ledge created by the knee and is centered there. You may have to bevel the end of the keelson for a snug fit at the transom. Clamp it in place.

4) Working forward, spring the rest of the keelson into the slots on the bulkheads and center mold. Clamp as you go.

STEM

PART #10

Figure 3-11. The stem is the front piece of the skeleton.

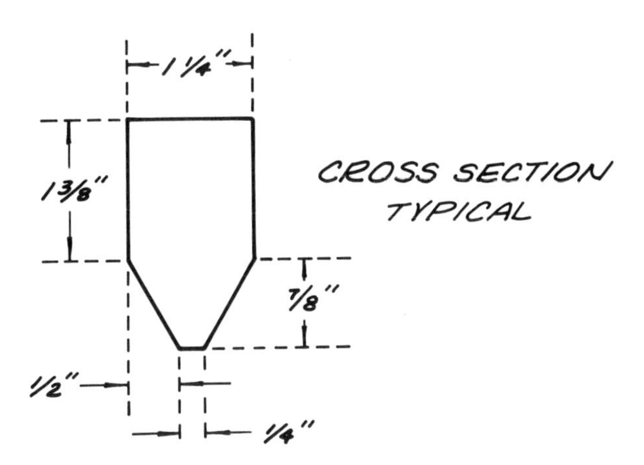

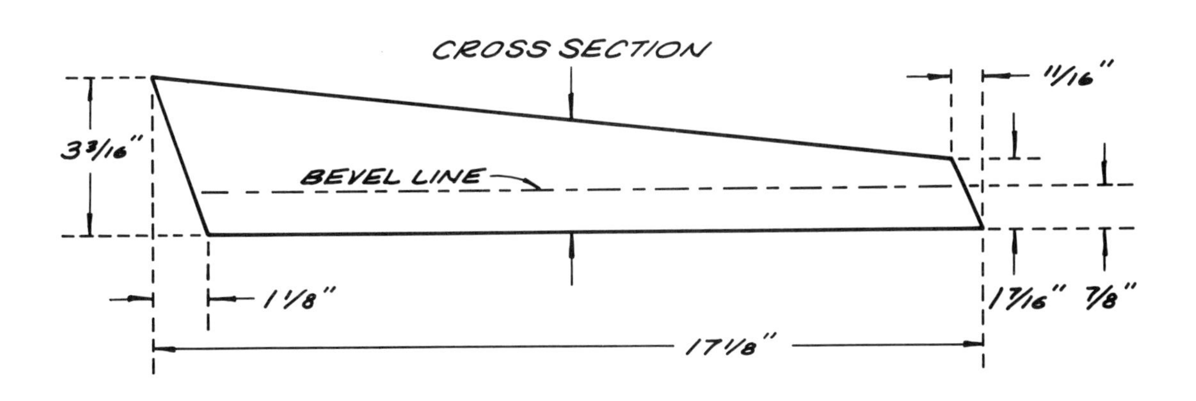

With any luck, everything lines up and the stem's centerline bends down and touches the centerline on the log. That's your goal – a good lineup stem to stern along the keelson. You may have to widen some of the notches along the way. (Check out Figure 3-12.)

5) Hold the stem tip in place (and the whole keelson, if necessary) by tying it down to the ladder with twine. When everything is in line and fits, drill a pilot hole through the lower tip of the stem at an angle (from the outboard side) into the log underneath. A vertical centerline drawn down the forward edge of the stem should meet the horizontal centerline on the log.

6) With the keelson firmly in place all along the jig, leave everything in suspension. You are ready to assemble the daggerboard case.

Figure 3-12. The stem is attached to the keelson and to the log.

PART 11: DAGGERBOARD CASE

The daggerboard case is a box through which the daggerboard is inserted after Clancy sets out from shore. Once assembled, the case is attached to the keelson.

Ingredients

- 1/2-inch mahogany marine plywood (sides of case)
- Six 1 × 2-inch clear fir strips, each 14 inches long
- Bronze screws (3/4 inch long)
- Carpenter's glue
- Epoxy resin and hardener
- Wood flour
- Marine varnish

There are some ingredients on our list that you might be using for the first time, such as epoxy and wood flour. They are described in the directions that follow. You will be using them again and again in later stages of our recipe.

Directions

1) Consult the pattern in Figure 3-13 and cut out the two daggerboard case sides, labeling them "top" and "bottom," "forward" and "aft" as shown.

2) Cut out six fir strips, each 14 inches long, to be trimmed to exact size later.

3) Give all inside surfaces of the box you are building three full coats of marine varnish.

4) On the outside surfaces of each side, flush with the top and bottom edges, lay the strips. The 3/4-inch (narrower) edge of the strip touches the side. Fasten temporarily with screws driven through the sides into the strips. Check that all the edges are flush. Trim the strips to fit and sand smooth. Then remove the screws. Apply glue and screw the strips back on.

5) The two remaining strips are sandwiched between

the two sides of the case (running flush, top to bottom) to create a 7/8-inch-wide opening to accept the daggerboard. This means you will have to saw these two strips lengthwise so that they are 7/8 inch wide (not 1 1/2 inches wide).

6) Mix up epoxy resin and hardener according to manufacturer's directions, then add plenty of wood flour until it reaches the consistency of thick cream.

Note: Both the epoxy resin and the wood flour are products that you can purchase at marine supply stores. The epoxy resin is often used in fiberglassing; it comes with a hardening agent that is added at the time you are ready to apply the epoxy. The hardened epoxy is then stiffened by mix-

Figure 3-13. Cut out two sides for the daggerboard case from this pattern and assemble as shown.

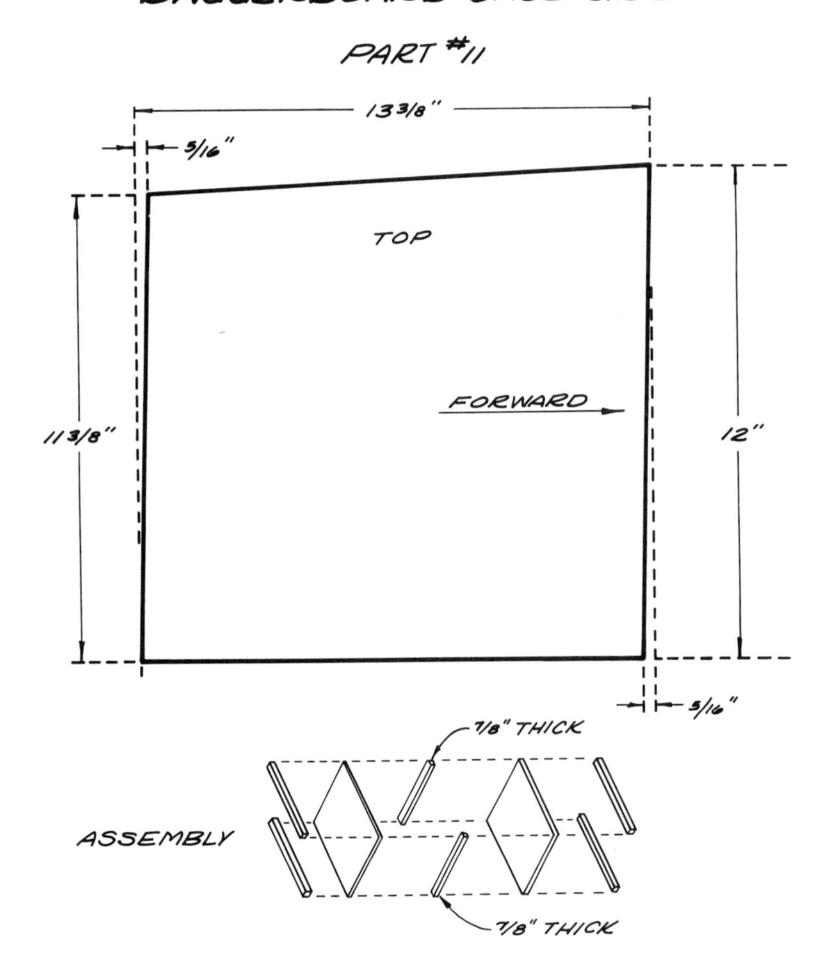

ing in the wood flour. A small plastic cup makes a good container for this concoction. Mix it up with a stick.

7) Spread the epoxy/wood flour putty on the strips for a watertight bond. Then screw through the sides into the strips. The case you are assembling should look like the one in Figure 3-14.

8) When the daggerboard case has cured a few hours, position it under the keelson in front of the forward bulk-

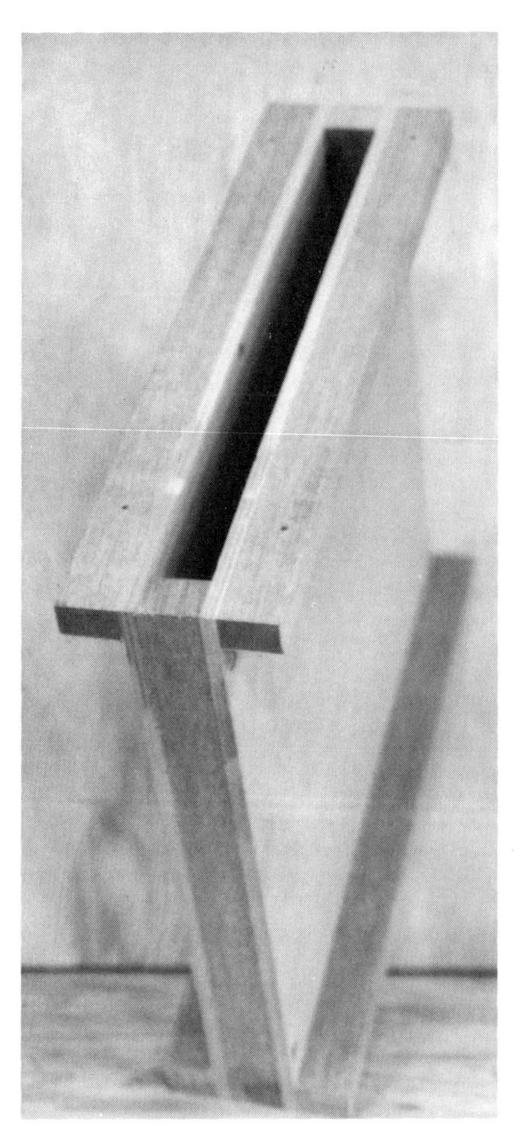

Figure 3-14. The daggerboard case resembles a long mail slot before being installed on the keelson.

Figure 3-15. The daggerboard case is positioned between the keelson and the forward bulkhead (looking aft*).*

head. Shove the *aft* face of the case against the *forward* face of the bulkhead. Clamp the daggerboard case to the underside of the keelson and let it hang (Figure 3-15).

CONNECTING PARTS OF THE SKELETON

In the Clancy version of an old popular song, we could say that the knee connects to the keelson, the keelson connects to the bulkhead, the bulkhead connects to the daggerboard case, the daggerboard case connects to the keelson, and the keelson connects to the stem.

Ingredients

- Pencil, tape, stick, and twine
- Marine adhesive sealant (no silicones)
- Bronze screws (1-inch × #8 and 1½-inch × #8)
- Epoxy resin and hardener
- Wood flour

Directions

1) Tape a pencil to a long stick. Poke it up inside the daggerboard case until it touches the underside of the keelson. Trace the case opening onto the keelson. Save the pencil-on-a-stick for later.

2) Unclamp and remove the daggerboard case. Drill pilot holes up through the keelson so that they mark the four corners of the opening you just traced.

3) Using the pilot holes, cut a daggerboard opening through the keelson. Clamp the daggerboard case back on the keelson, check the opening, and enlarge it if necessary.

4) Unclamp the daggerboard case, spread marine adhesive sealant on the top rim, and clamp it back on the keelson. (This adhesive should not contain any silicones.) Drive some 1-inch bronze screws, three on each side, down through the keelson into the daggerboard case. This is a watertight joint, so squeeze out excess sealant.

5) Free up the keelson at the stem. Spread thickened epoxy mix (containing hardener and wood flour) on the daggerboard case where it meets the bulkhead and on the bulkhead slots—*but not on the center mold or the center mold slot*!

6) Spring the keelson back into place; tie it down with twine or a bar clamp at the stem. Then free up the keelson at the transom knee, apply glue, and drive two 1½-inch × #8 bronze screws through the keelson into the transom knee. Countersink the screws. Tip: Using a countersink with a stop collar on your drill is a sure way to recess the screw heads evenly.

Figure 3-16. Under the watchful eye of this feline supervisor, the keelson in clamped into position.

7) Drive three 1-inch × #8 bronze screws forward through the forward bulkhead into the daggerboard case.

8) Drive a $1^1/_2$-inch bronze screw through the stem into the log.

Now sit back and allow everything to cure overnight. In the morning, untie and unclamp the skeleton. Not only is the jig up, but so are the keelson, bulkheads, daggerboard case, stem, and transom. In other words, you have passed through the two dullest stages of Clancy building. Hereafter, Clancy begins to look like a boat.

FOUR

The Hull Story and Nothing But

Clancy's hull consists of two bottom panels and two side panels, plus the keel. Hot-melt glue, bronze screws, and epoxy resin are the key adhesive ingredients; you'll also need to fiberglass some of the seams. (Don't worry if you've never used fiberglass before; if you've used scissors and a paint brush, you can fiberglass a boat).

In this chapter, we help you put the bottom and sides on first, then the keel. After coating the seams of the hull with fiberglass, you can remove Clancy from the jig.

PART 12: BOTTOM

Clancy's bottom consists of two identical panels joined on the keelson.

Ingredients

- $1/4$-inch 5-ply mahogany or fir marine plywood, 4 × 10 feet
- Bronze screws ($3/4$-inch × #6)
- Marine adhesive sealant (no silicones)

Directions

1) Cut out two identical bottom panels following the dimensions given on the pattern (Figure 4-1).

2) Clamp both panels on top of the jig, one at a time. Secure these bottom pieces temporarily with screws driven along the keelson and at the transom. Screws placed in the

TOOLBOX

Clamps (spring clamps, C-clamps)
Drill (reversible) with countersink and stop collar
Drill bits
Hammer
Hot-melt glue gun
Paint brush
Pencil
Pencil-on-a-stick
Plane
Putty knife
Saber saw
Scissors
Scraper
Speedblock sander
Squeegee
Tape measure
Wood rasp

transom must be driven at an angle. Use care here: screw tips must not penetrate the outside face of the transom and they must be started far enough from the edge so that they can be covered later. Note: The panels will be slightly oversize, but do not trim them yet.

3) Remember the pencil-on-a-stick trick you performed with the daggerboard case? Here's where you need to trace the daggerboard opening on the bottom panels. When you are done, keep this tool handy.

4) Remove the bottom panels and cut out the daggerboard slot you traced there.

5) Squeeze out a bead of adhesive sealant along the keelson and edge of the transom; spread that bead with a putty knife.

6) Position the bottom panels, one at a time, so that you can use the screw holes drilled previously. Secure the panels with bronze screws. Place a screw every 6 inches or so along the keelson, the two bulkheads, and the transom. Work

Figure 4-1. Cut out two identical bottom panels.

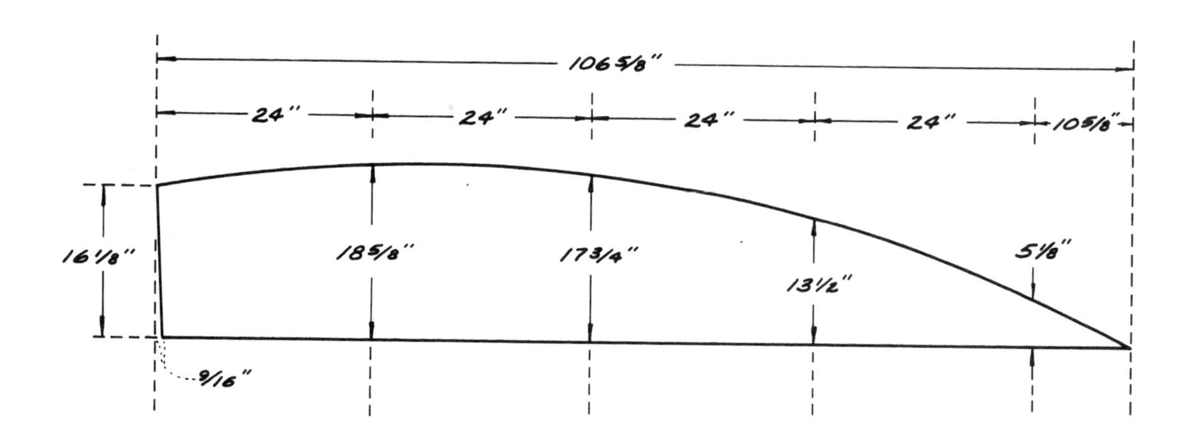

outward from the center of the boat. At the transom, remember to angle the screws so they will be properly aligned (Figures 4-2 and 4-3 show the bottom panels on).

7) Set all screw heads deep enough so that you can

Figure 4-2. Position the bottom panels on the skeleton.

Figure 4-3. Attach the bottom panels only after cutting a slot for the daggerboard.

putty over them later. Tip: Using a countersink with a stop collar on your drill is a sure way to recess the heads evenly.

8) Trim the edges of the bottom panels slightly (using a plane or similar tool) so that they are flush with the outboard edges of the bulkheads, center mold, stem, and transom.

PART 13: SIDES

The two identical side panels require quite a bend. They overlap the edge of the bottom and are trimmed later. Since each piece of plywood has its own idea of how to bend, you'll welcome another pair of hands when fitting the sides. By the way, you have now come to the heart of the glue-and-stitch method, so get out your glue gun.

Ingredients

- 1/4-inch 5-ply marine plywood, 4 × 10 feet, mahogany or fir
- Bronze screws (3/4-inch × #6)
- Epoxy resin and hardener
- Wood flour
- Hot-melt glue sticks

Figure 4-4. Cut out two identical side panels.

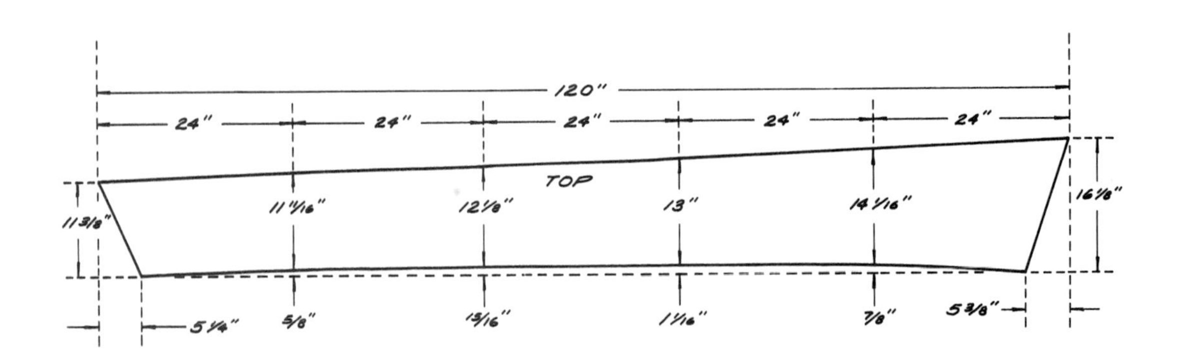

SIDE

PART #13

Directions

1) Cut out two identical side panels from the pattern dimensions (Figure 4-4).

2) Make a dry run with each side panel to be sure it fits (overlapping the bottom, transom, and stem).

3) Thicken up some epoxy resin by mixing in hardener, then wood flour. Butter the edges of the transom and stem (but not the center mold).

4) Temporarily clamp or screw the side panel to the transom and stem (with a slight overlap). See Figure 4-5.

5) Heat up the hot-melt glue gun. Stitch the side in place by shooting a bead of glue every foot as you go and pressing in for a minute so the glue can set there. See Figure 4-6. You are gluing and pressing where the bottom edge meets the side, and this is where a friend's hands are indeed handy.

6) Every 6 inches or so drive a 3/4-inch × #6 bronze screw through the side into (a) the transom and (b) the stem.

Figure 4-5. Temporarily clamp the side panels in position.

Figure 4-6. The stitch-and-glue method: Secure the side panels with hot-melt glue and then drive the screws.

Do not drive screws through the side into the forward and aft bulkheads or into the center mold.

7) Set all screw heads deep enough so that you can putty over them later. A countersink and stop collar attached to your drill makes this easy.

PART 14: KEEL

The keel, a mirror image of the keelson, covers the gap you have now created between the two bottom panels.

Ingredients

- 1 × 4 fir
- Bronze screws (3/4-inch × #6)
- Epoxy resin and hardener
- Wood flour
- Finishing nails (1 inch long)

Directions

1) Cut out the keel according to the pattern (Figure 4-7).

2) Temporarily secure the keel in place along the gap between the bottom panels with a nail at each end.

3) Repeat the old pencil-on-a-stick trick with the daggerboard case yet another time—tracing the case opening on the underside of the keel.

4) Remove the keel and cut out the daggerboard slot.

5) Mark on the keel the positions of the screws holding the bottom so that you don't hit those screw heads when you fasten the keel over them.

6) Spread a mixture of epoxy resin, thickened with

Figure 4-7. The keel, with the daggerboard slot located and cut, covers the gap where the bottom panels join along the keelson.

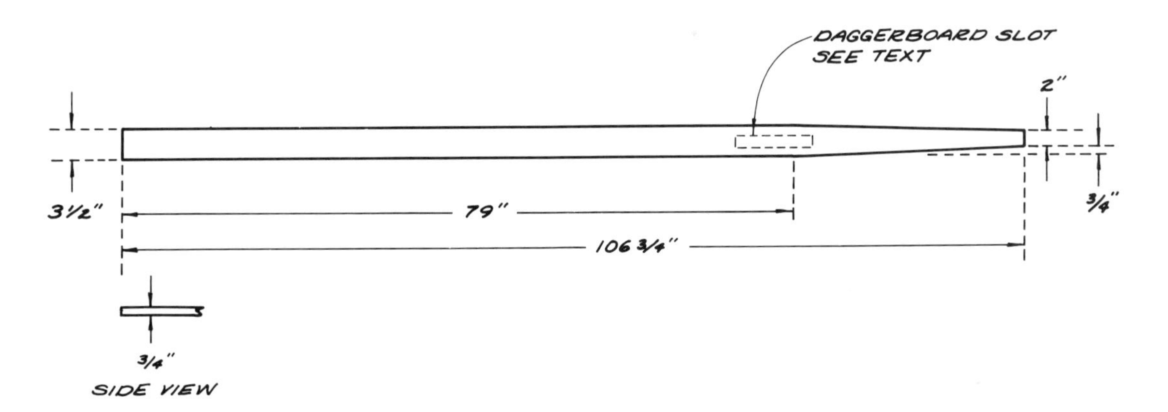

Figure 4-8. Attach the keel to the bottom.

wood flour to the consistency of mayonnaise, along the gap between the bottom panels, stem to stern.

7) Position the keel and secure it with screws every 6 inches on either side of the centerline. Countersink all screw heads deep enough to putty over (Figure 4-8).

8) Plane and sand the tip of the keel at the transom end so that it conforms to the angle of the transom.

FINISHING THE HULL

You have completed construction of the hull, but it still requires your finishing touch. You need to fair it, then tape the seams.

Fairing the hull means filling the screw holes with putty and rounding all the edges with a scraper and sandpaper.

Taping the seams means applying strips of fiberglass tape where Clancy's pieces meet and coating them with epoxy.

The result of fairing and taping is a smooth, watertight hull.

Ingredients

- Epoxy resin and hardener
- Wood flour
- Barrier cream
- Plastic gloves
- Safety goggles
- Scissors, paint brush, squeegee, towels
- Pie tin or plastic cups
- Fiberglass tape (3-inch-wide roll)
- Scraper, plane, sander (speedblock)

Directions for Fairing

1) Following the manufacturer's recipe, mix up some epoxy resin and hardener. Stir in enough wood flour to produce a concoction as thick as peanut butter.

2) With a putty knife, patch over all the screw heads (as in Figure 4-9). Wipe up the excess with a towel.

3) Then, when the putty is cured, but still fresh, scrape or sand the patches smooth.

4) Using a block plane, a surform, or a wood rasp, trim the side panels where they overlap the bottom, the stem, and the transom (see Figure 4-10).

5) With the same trimming tools, round off all the corners and sides of the hull, creating a 1/4-inch radius (curvature).

Directions for Taping

First, a reminder: *Always maintain a protective screen between yourself and epoxy resins and fiberglass by applying barrier cream to your hands and arms, or wearing plastic gloves. Use safety goggles. Thus suited up for battle, follow this sequence:*

1) From a roll of 3-inch-wide fiberglass tape, cut two strips to cover each seam where wood meets wood: at the transom, at the stem, where the bottom meets the sides (the

Figure 4-9. Patch the screw heads with putty.

Figure 4-10. Fair the hull with a scraper or speedblock sander.

chines). The tape overlaps for extra strength and shouldn't be stretched, so use three double strips (not one) to do the three sides of the transom.

2) Drape the tape (2 strips for each seam) over the hull near the spots it will be used (Figure 4-11). Mix about 6 ounces of epoxy resin and hardening agent at a time in a cup (Figure 4-12). Tip: If you pour this mixture into a shallow pie tin, you will extend its working life.

3) Lay the tape over a seam, centering it. Using a paint brush or squeegee, tack the tape on with some globs of resin. Quickly saturate the whole strip of tape with resin, work out air bubbles with the squeegee, and wipe up drips with a towel as you go (Figure 4-13).

To speed up the taping, here's our favorite technique: Apply a second strip of tape while the first is dripping wet. Place it on the first layer, but overlap it slightly at the ends so that the two layers don't begin and end in exactly the same place on the seam. Do the second layer just as you did the first layer (Figure 4-14).

4) Be sure to catch all drips as you go, before they harden. Squeegee out all air bubbles, too.

5) Before the resin is rock hard (within 6 to 24 hours) use a constantly sharpened paint scraper to knock off all the bumps where excess resin has gathered. Scrape down the edges of the tape, too.

6) When all seams are twice-taped, twice-wetted, and at least once-scraped, you are done taping. Except for this: You must feather all the tape edges with a scraper and sandpaper. Feathering means to smooth the edges so that they blend into the surrounding wood.

Make Clancy smooth to the touch and fair to the eye. Then you are really done taping, unless you used fir instead of mahogany marine plywood for the hull.

If You Used Fir

If you built the hull of fir—rather than mahogany plywood—you must now sheathe the entire hull in fiberglass cloth to prevent *checking* (the plywood's surface will crack and splinter due to shrinkage). The process is similar to taping. Use the same epoxy mixture and tools. The difference is that now you must apply and coat a wide piece of fiberglass cloth from a bolt. Here's how to do it:

1) Drape a sheet of lightweight (6-ounce) fiberglass cloth (60 inches wide) over the bottom of Clancy's hull, overlapping the sides and transom by 2 inches.

Figure 4-11. Lay out two strips of tape for each seam on the hull.

Figure 4-12. Mix epoxy resin and hardener in a cup, about six ounces at a time.

Figure 4-13. Apply epoxy mixture to the taped seams.

Figure 4-14. Quickly lay a second strip of cloth tape over the first strip, wet with epoxy, and squeeze out the excess with a squeegee.

2) Pour some epoxy mixture on the cloth, spreading it with a brush and squeegee. Work from the center outward, chasing away the wrinkles. If some wrinkles remain, they can be scraped and sanded away later. Also mash out the air bubbles that develop. They are a little harder to see than they were in the tape. The cloth surface may seem a little fuzzy, but don't worry–and don't add more resin (it will only add weight, not strength).

3) Repeat this same exact process (steps 1 and 2) on the sides (and the bottom, too, if you used fir for that piece), overlapping the cloth 2 inches over the bottom and across the aft face of the transom.

4) Do a final sanding and feathering.

5) Apply a coat of epoxy mixture to the entire hull. When it sets up, wipe it off with a weak ammonia and water solution. The fuzzy look should be about gone.

6) Apply a second coat of epoxy mixture (it fills any remaining pinholes). Wipe off again with ammonia and water.

7) Sand the hull smooth.

That's all, although some boatbuilders like to do two coats of cloth (sanding and wiping down between coats). Each coat Clancy dons, by the way, adds another five pounds.

UNJIGGING CLANCY

Time to turn Clancy loose from the jig and right side up so that you can work on the inside.

Directions

1) Find a friend to give you a hand.

2) Tap the hull loose from the jig. If you are lucky, Clancy is not stuck to the center mold and crutch. The bulkheads should slip out of the notches, too.

3) With one of you on each end, carefully but firmly wrench the hull off the jig and lift it up free.

4) Turn the hull over and set Clancy down to one side (Figure 4-15).

Figure 4-15. Clancy unjigged!

Voilà! You are done with the jig. Take the jig off the saw-horses and replace it with Clancy, right side up, resting on the keel at a good working height.

You can store the jig as is or disassembled for a head start on the next Clancy.

GLUE THE BULKHEADS

One final step. There are some loose cannons aboard. As soon as Clancy is off the jig, be sure to glue the forward and aft bulkheads (Parts 5 and 6) to the bottom and sides. Only three steps are required:

1) Square each bulkhead to the keelson in the bottom.

2) With your hot-melt glue gun at your side, push down on the bottom under each bulkhead and shoot spots of glue into the open seam.

3) Spread the sides and spot again with hot glue.

This will hold the bulkheads in position until later, when you can really fasten them down with putty, epoxy, and fiberglass tape. But that's another chapter. For now, you can admire what you have built. Free from the jig, Clancy is beginning to look like a boat.

FIVE

Inside Clancy:
Of Kingplanks and Carlins

There are 10 new pieces to add to the interior. Most of these pieces serve as support for the wraparound deck. They are numbered as follows:

Part 15: Forward Bulkhead Doubler

Part 16: Aft Bulkhead Doubler

Part 17: Partial Bulkheads

Part 18: Carlins

Part 19: Deck Beam

Part 20: Forward Kingplank

Part 21: Aft Kingplank

Part 22: Guardrails

Part 23: Cutwater

Part 24: Mast Step

Figure 5-1 gives you an exploded view of how these pieces fit together (except the mast step, which is added last). In addition, we give you instructions in this chapter for filleting and taping the interior seams so that Clancy remains lightweight yet extremely strong.

TOOLBOX

- Clamps (spring clamps and C-clamps)
- Drill (reversible) with countersink and stop collar
- Drill bits
- Handsaw
- Paint brush
- Pencil
- Pencil-on-a-stick
- Plane
- Putty knife
- Saber saw
- Scissors
- Scraper
- Shoe file
- Speedblock sander
- Square
- Squeegee
- Tape measure
- Wooden dowel
- Wood rasp

PART 15: FORWARD BULKHEAD DOUBLER
PART 16: AFT BULKHEAD DOUBLER

The bulkhead doublers are support strips added to the forward edge of the forward bulkhead and the aft edge of the aft bulkhead. They stiffen the boat and provide additional deck support.

Ingredients

- 1 × 4 fir or mahogany or scrap lumber
- Bronze screws (1 inch or less in length)
- Carpenter's glue

Figure 5-1. Exploded view of the interior and deck.

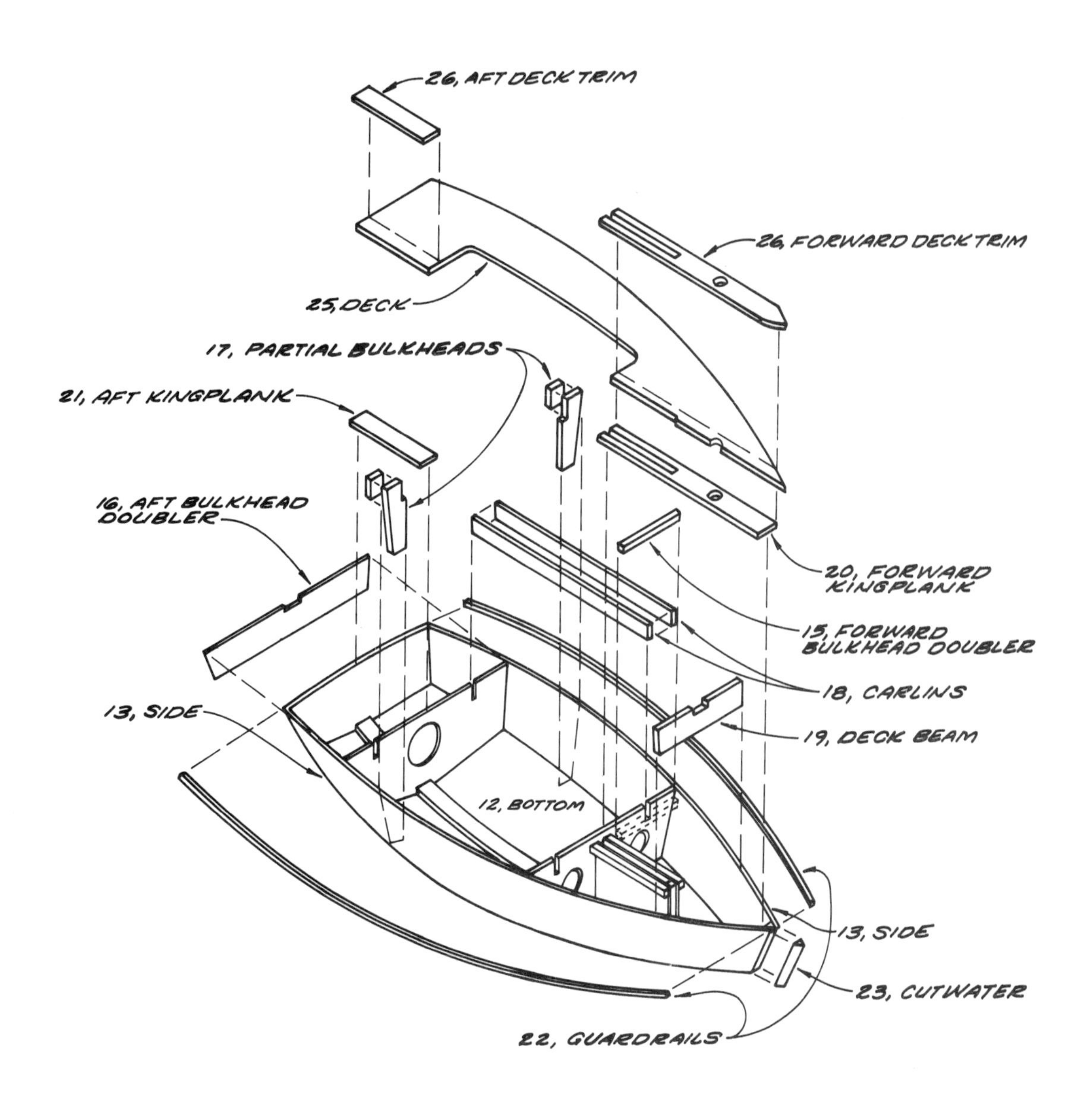

Directions

1) Cut out two doublers for the forward bulkhead doubler (Part 15: Figure 5-2).

2) Cut out one doubler for the aft bulkhead doubler (Part 16: Figure 5-3).

Figure 5-2. Cut out two forward bulkhead doublers.

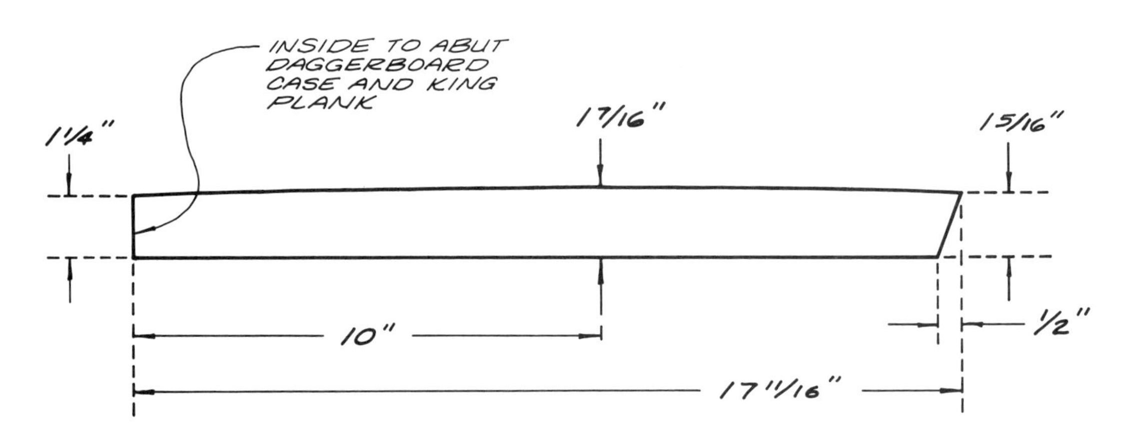

FORWARD BULKHEAD DOUBLER

PART #15

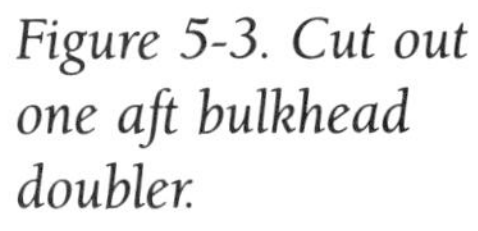

Figure 5-3. Cut out one aft bulkhead doubler.

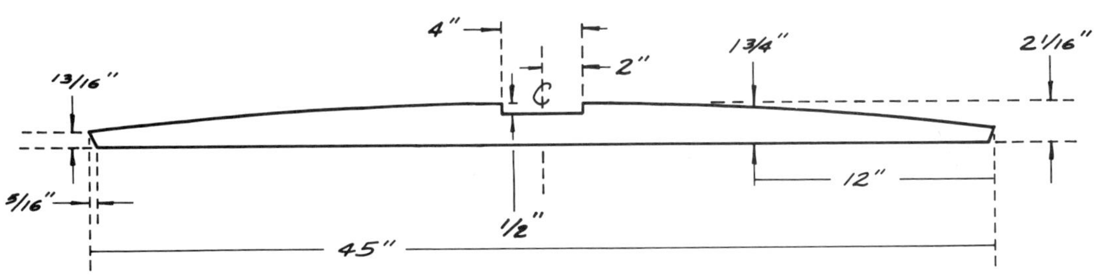

AFT BULKHEAD DOUBLER

PART #16

3) Attach the two doublers to the forward bulkhead on the forward side (the daggerboard case side) using glue and screws driven forward through the back side of the bulkhead.

4) Attach the single doubler to the aft bulkhead on the aft side (the transom side) using glue and screws driven aftward through the bulkhead into the doubler.

The top edge of the doublers should be even with the top edge of the bulkheads.

PART 17: PARTIAL BULKHEADS

The partial bulkheads are installed on either side of Clancy and support the wooden strips known as carlins (which you will install next).

Ingredients

- 1/4-inch 5-ply mahogany or fir marine plywood (partial bulkheads)
- 2 × 2 clear fir (bulkhead blocking)
- 1 × 4 fir or mahogany strips, scrap is okay (partial bulkhead doublers)
- Bronze screws (3/4-inch × #6)
- Carpenter's glue

Directions

1) Cut out two partial bulkheads from our pattern (Figure 5-4).

2) Cut out two blocking pieces to fit back of the notch (also shown in Figure 5-4).

3) Glue the blocking to each partial bulkhead on the *aft* face running flush with the notch for the carlin. Secure with two 3/4-inch × #6 bronze screws.

4) Cut out two doubler strips, each less than 9 inches long. These are attached on the *forward* face of each partial bulkhead so as to run horizontally from the notch to the side of the boat. The doublers thus provide additional deck support later. Fasten each doubler with glue and 3/4-inch × #6

bronze screws driven through the partial bulkhead into the doubler.

5) Position the partial bulkheads opposite each other about halfway between the bigger forward and aft bulkheads (Figure 5-5). Shape the partial bulkheads to fit snugly against the sides. Be sure the partial bulkheads are square up and down against the sides.

6) Secure the partial bulkheads against the hull sides using glue (see Figure 5-6). A scrap stick laid across the boat will help keep the partial bulkheads lined up while the glue is setting.

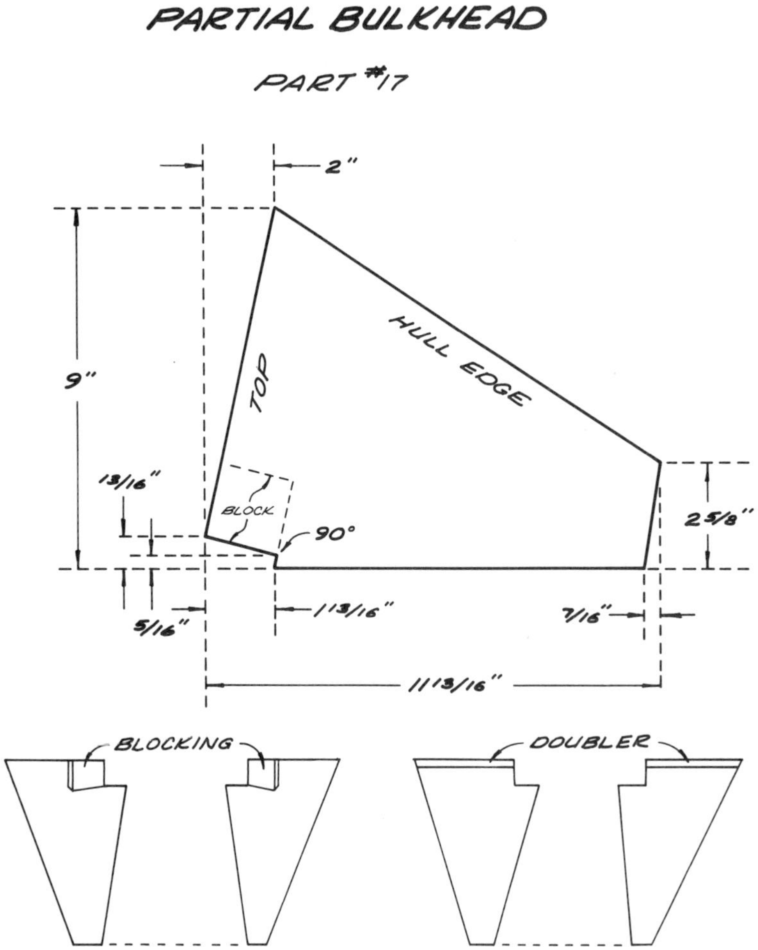

Figure 5-4. Cut out two partial bulkheads and attach blocking and doublers as shown.

Figure 5-5. Locate the partial bulkheads midway between the forward and aft bulkheads.

Figure 5-6. Glue the partial bulkheads in place against the hull.

PART 18: CARLINS

The carlins, which rest in the partial bulkheads you just installed, support portions of the deck on either side of the cockpit. The tips of the carlins are inserted in the notches in the full bulkheads forward and aft.

Each of the two carlins is made up of two identical strips laminated (glued) together for strength.

Ingredients

- 4 strips of 1/4-inch 5-ply mahogany or fir marine plywood, 2 inches wide and 48 3/8 inches long
- Bronze screws (1-inch × #8)
- Epoxy resin and hardener
- Wood flour

Directions

1) Cut out four identical carlin pieces from the carlin pattern (Figure 5-7). Each of the two carlins is made up of two pieces. They will be glued together in place in the boat. The carlins fit in the slots in the forward and aft bulkheads and in the notch in the partial bulkheads. Measure so that the

Figure 5-7. Cut out four pieces, gluing two together to form each of two carlins.

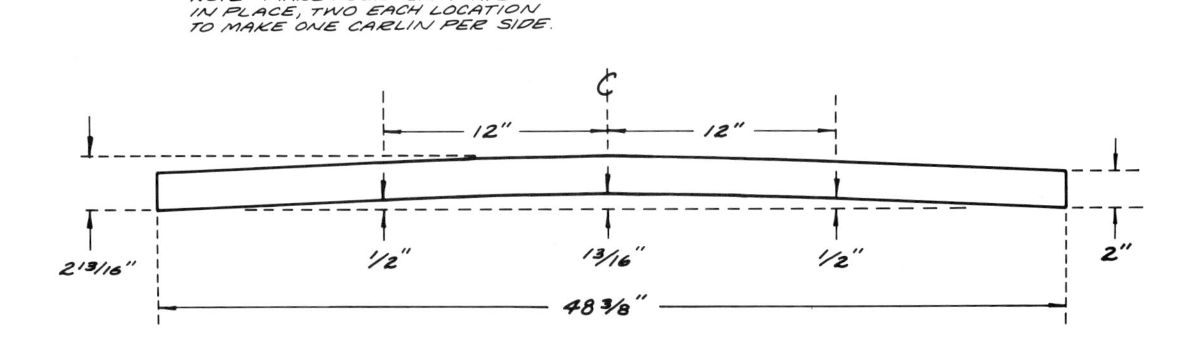

carlins will fit exactly in your Clancy, cut them out, try them in place, and trim to fit.

2) Mix up some epoxy to the consistency of stiff heavy cream.

3) Work with one set of carlins on one side of Clancy at a time. Butter a slot in the forward bulkhead, a slot in the aft bulkhead, and a corresponding notch in one partial bulkhead, spreading the thickened epoxy on these three ledges.

4) Butter the entire side of one carlin piece (where it will be sandwiched face to face with the other carlin piece) and spring it into place on the bulkheads. You've just put half a carlin in place.

5) Quickly spring the other half of the sandwich (a second carlin piece) into position, too, so that it presses against the thickened epoxy/wood flour spread. Drive two screws through this newly laminated carlin into the blocking on the partial bulkhead to hold it in place (see Figure 5-8).

Figure 5-8. Position each carlin in the bulkhead slots and partial bulkhead notches, two strips on each side (glued together). Screw the carlins to the blocking on the partial bulkhead notch.

Figure 5-9. Clamp the carlins in place so the glue can set up overnight.

6) Since you are laminating the carlin together while in place in the boat, you will need to use plenty of C-clamps or spring clamps or both to keep the sandwich closed tight while the thickened epoxy cures (Figure 5-9). So unless you have as many clamps as Imelda Marcos had shoes, we recommend that you put in the carlins one set at a time.

7) Trim the ends of the carlins flush using a handsaw.

8) Work the lower horizontal edges of both carlins, sanding and rounding them smooth.

FILLETING TOOL

Before continuing interior work, we recommend that you make your own filleting tool from our pattern (Figure 5-10).

Cut it from a 1/16-inch piece of stiff but flexible plastic or from an existing squeegee.

You'll use your filleting tool to lay down a cove of putty along the seams where pieces meet inside.

FILLETING CLANCY

Filleting Clancy simply means coating the interior seams with a curved bead of putty called a cove. The cove-shape of the putty enables you to lay fiberglass tape in the corner seams smoothly, with no voids. When a corner seam is taped on a cove of putty, the strains are evenly distributed and stress points don't develop there.

We've found that it's fastest to fillet and tape each corner seam as a single process. While the putty cove is still wet, we lay cloth tape on it and apply epoxy. If you want to do the same thing, *read all the directions* that follow on filleting and taping–then decide. You can do each step separately or combine them.

Directions for Filleting

1) Whip up a fairly stiff mixture of epoxy resin, hardener, and wood flour–about 6 ounces worth in a cup.

Figure 5-10. Cut a filleting tool from 1/16" plastic or from a squeegee.

2) Apply this putty with a putty knife or (better) the plastic filleting tool you made up earlier in this chapter. You apply putty to almost every interior seam (where one piece meets another) and you shape it in a hollow, concave manner.

3) Don't fillet the following interior seams:

- Where the keelson meets the bottom
- Where the stem meets the keelson or the sides
- Where the daggerboard case meets the keelson or the forward bulkhead

4) You can allow the putty to harden and then give it a thorough sanding with 50-, 60-, or 80-grit sandpaper wrapped around a wooden dowel (Figure 5-11).

Directions for Taping

1) Each fillet needs to be taped. Cut two lengths of fiberglass tape to fit each coved seam. Allow some extra length. The tape can be trimmed easily with scissors even after it's wetted. Station the tape strips near their fillet (Figure 5-12).

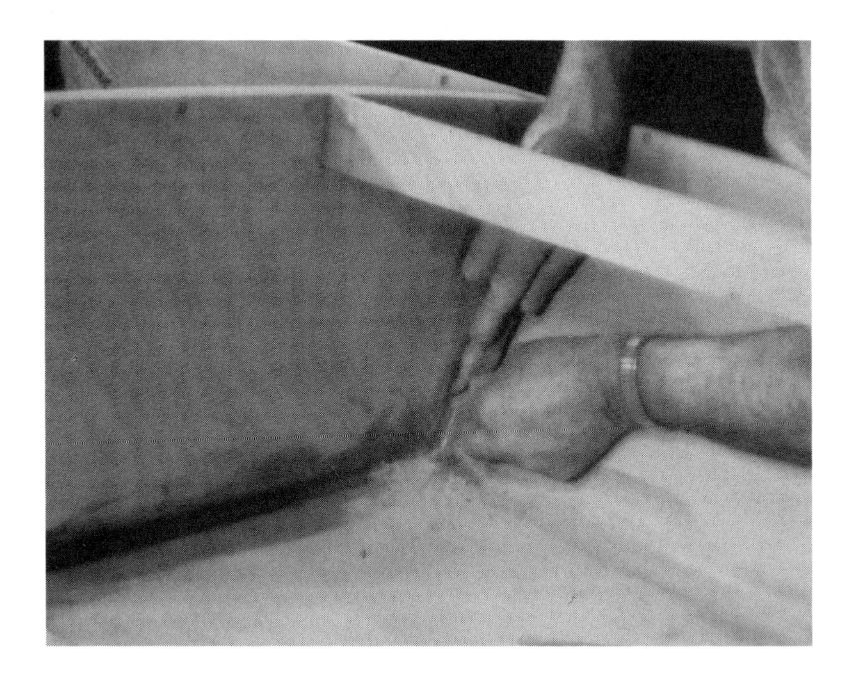

Figure 5-11. If putty coves have set up and hardened, sand and scrape them smooth before taping.

2) Brush the filleted seam with a mixture of epoxy resin and hardener.

3) Lay the cloth on the wet seam and brush with epoxy mixture (Figures 5-13 and 5-14).

4) Immediately apply a second layer of cloth to the

Figure 5-12. Station two layers of tape near each cove.

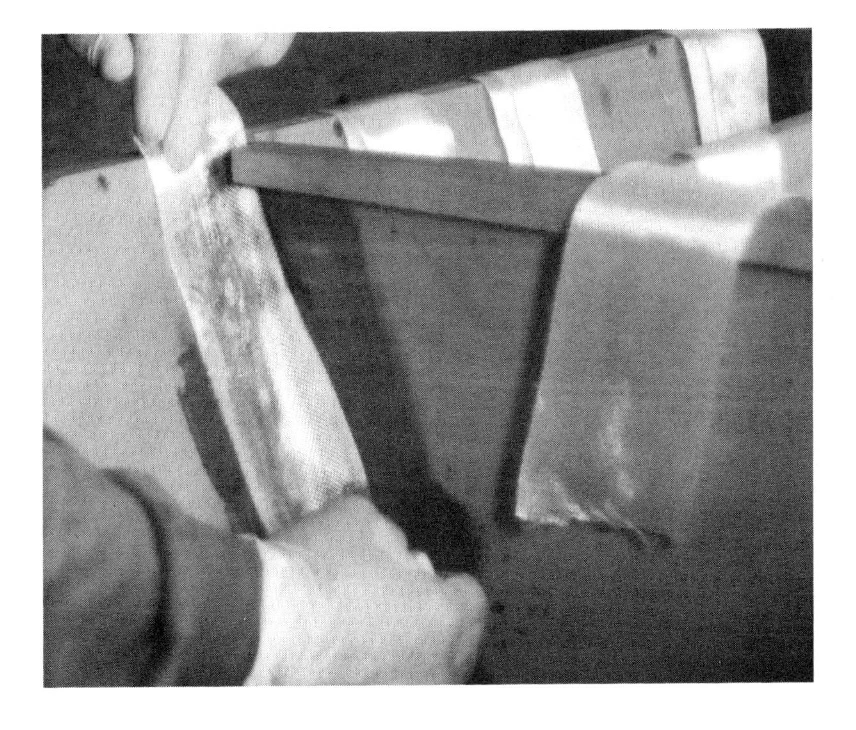

Figure 5-13. Wet the cove with epoxy and cover with tape.

same cove, wetting it, working out bubbles with a brush, trimming excess tape with scissors, and cleaning up excess epoxy with a towel (Figure 5-15).

5) Let the tape cure overnight.

Figure 5-14. Brush the taped cove with epoxy mixture.

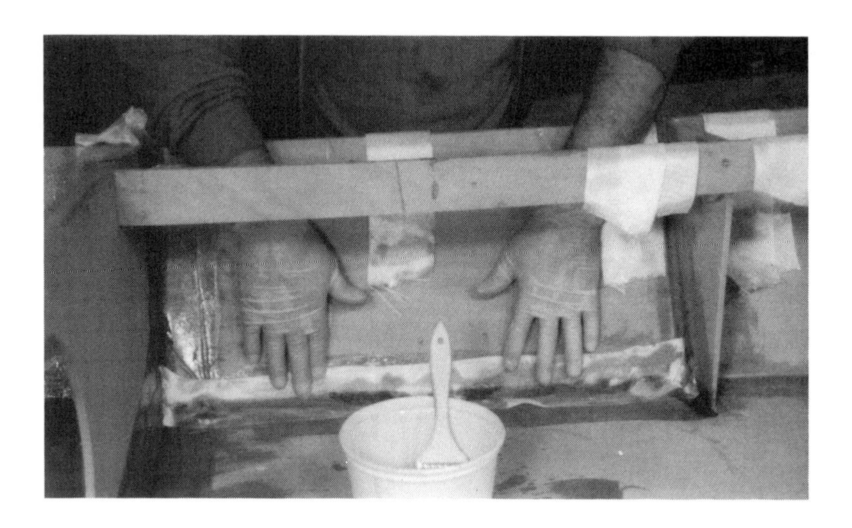

Figure 5-15. Immediately apply a second layer of tape and brush on the epoxy mixture.

Directions for Combining Filleting and Taping

If you want to speed through this process, skip the sanding (Step 4 in Filleting Directions) and while the putty is still wet apply fiberglass tape strips to the coves, wetting the tape with epoxy and using the filleting tool to smooth the putty as you go.

To combine the putty coving and the taping, you will need to plan well ahead. Set out all your supplies for fiberglassing first, and cut lengths of tape (double lengths, because you want to put two layers of tape on the interior) for all the coves. Have your epoxy mix and containers nearby. Mix up the putty first. As soon as you cove one or two seams, switch to taping. An assistant is invaluable if you decide to do two tasks at once.

If You Used Fir

If the bottom of your Clancy is made of *fir* rather than *mahogany* marine plywood, then you must also add one layer of fiberglass cloth to the cockpit floor at this time to prevent checking (just as you did to the bottom on the outside of the hull earlier).

Cut two sections of fiberglass cloth, one to cover the floor on each side of the keelson (between bulkheads). Apply epoxy mix as before and allow it to cure overnight. Don't sheathe the keelson.

FAIRING CLANCY

Take a very sharp scraper and scrape off the high spots where the epoxy is thick or has dripped, wherever the cloth is wrinkled and cries out to be feathered.

Then sand, sand, sand, and sand again. Sand until you reach a Zen-like state of pure contemplation or complete idiocy (as in Figure 5-16).

Tip: A shoe file is an excellent tool for feathering and scraping fiberglass tape.

Keep sanding with 40-grit paper until a feather edge is achieved. Switch to 80-grit paper for the final finish.

Now you're done: you've reached sanding nirvana.

Figure 5-16. Sanding as a form of enlightenment.

PART 19: DECK BEAM

The deck beam is a supporting crosspiece installed up front between the daggerboard case and the stem. It is held in place by filleting and taping–by the same process you just completed to reinforce other interior pieces (so this should be a snap to install).

Ingredients

- $^1/_2$-inch 5-ply mahogany or fir marine plywood, 23 inches long
- Hot-melt glue sticks
- Epoxy resin and hardener
- Wood flour

Directions

1) Cut out the deck beam using the pattern in Figure 5-17. Position it about midway between the stem and the forward bulkhead. The deck beam is a crosspiece which will support the deck up front (as well as the kingplank which crosses it).

2) Glue the ends of the deck beam to the sides of the hull. The top edge of the deck beam is flush with the top edge of the sides (as in Figure 5-18).

3) When the glue has hardened a bit, prepare some epoxy/wood flour putty and apply it in a thick bead up and down on both sides of the joint between deck beam and hull. Use your filleting tool to give the putty its cove-like shape.

4) Apply two layers of tape to the putty cove. Sand fair.

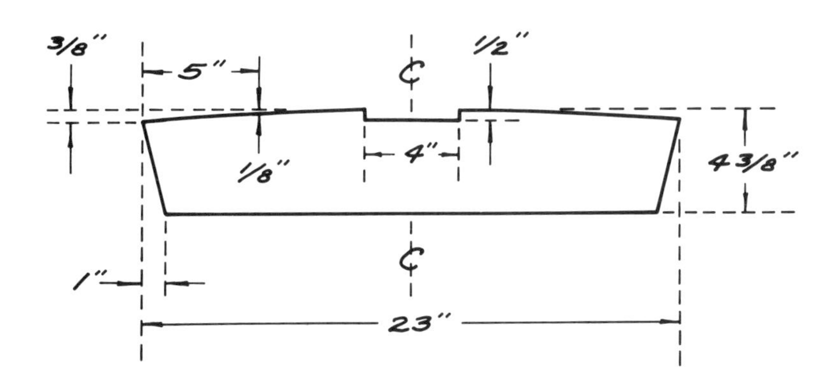

Figure 5-17. The deck beam is a supporting crosspiece installed between the daggerboard case and the stem.

Figure 5-18. Glue the deck beam in position midway between the stem and forward bulkhead.

PART 20: FORWARD KINGPLANK
PART 21: AFT KINGPLANK

The kingplanks, forward and aft, provide stiffness and decking support.

Ingredients

- 1/2-inch 5-ply mahogany or fir marine plywood, 4 inches wide (for kingplanks)
- 1/2-inch or 3/4-inch scrap fir or plywood, 4 inches wide and 2 inches long (for transom block); 3 inches long (for stem block)
- Bronze screws (1-inch × #8)
- Carpenter's glue
- Epoxy resin and hardener
- Wood flour

Directions

1) Cut out both kingplanks from the patterns (Figures 5-19 and 5-20). Determine the exact length of each piece by measuring what is required in your Clancy. (In the mysterious business of boatbuilding, distances can vary from boat to boat.)

2) Position the aft kingplank. It should be crammed

Figure 5-19. The forward kingplank supports the front of the deck.

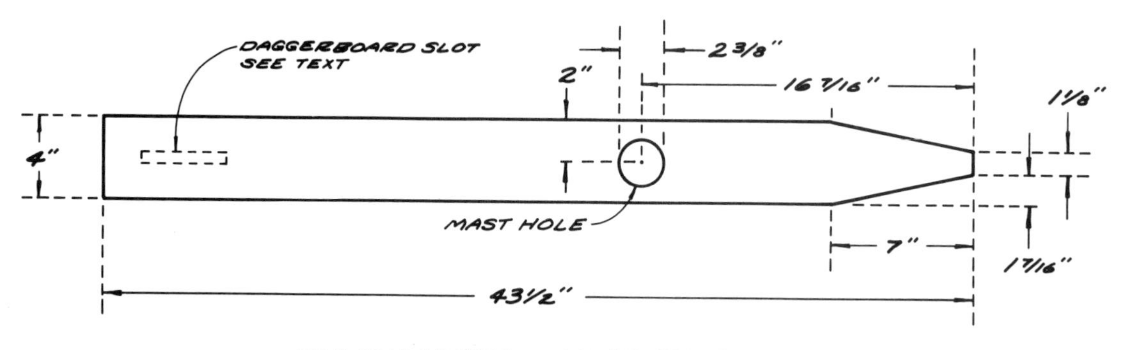

snugly between transom and bulkhead, flush with the top of each. The after end must be beveled to meet the slant of the transom.

3) Mark where the underside of the aft kingplank tip touches the transom. Cut out the transom block and attach it as a ledge where you marked, using glue and two screws (driven through the transom).

4) Glue the aft kingplank to the transom support block and the slot in the aft bulkhead doubler. Screw downward into the block and doubler on each end. Correct attachment is shown in Figure 5-21.

5) Now position the forward kingplank and trim for a snug fit. Once again you must use the old pencil-on-a-stick trick to trace out the opening of the daggerboard case (although you could just cut a 1-inch-wide slot down the center of the kingplank, stopping at the forward edge of the daggerboard opening).

6) Using a saber saw, cut a 2³/₈-inch hole for the mast at the position indicated on the pattern (in Figure 5-19).

7) Glue and screw a support block on the inside face of

Figure 5-20. The aft kingplank supports the back of the deck.

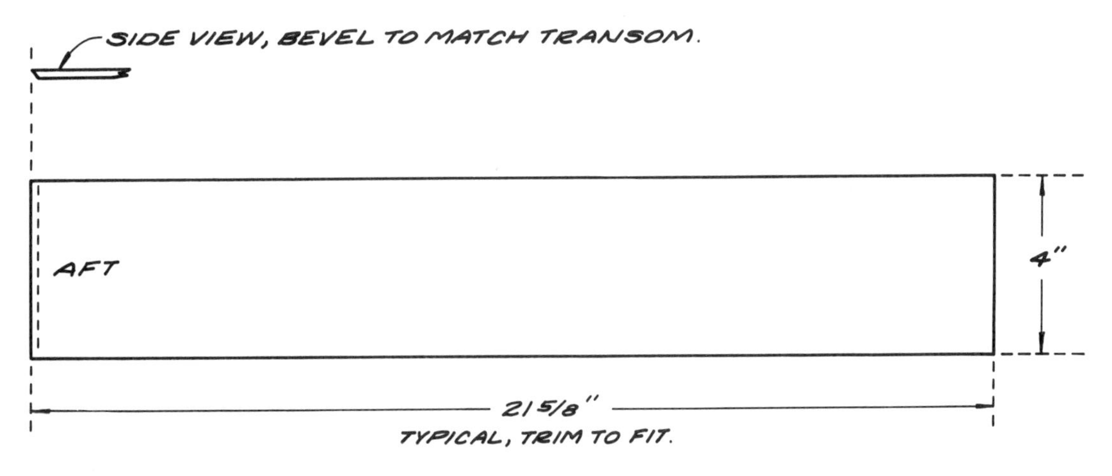

Figure 5-21. Position the aft kingplank in the ledges on the transom and aft bulkhead doubler.

the stem so it acts as a ledge to accept the forward tip of the forward kingplank.

8) Finally, spread epoxy/wood flour putty on all contact surfaces (including the stem area). Drive two screws through the forward bulkhead into the end grain of the kingplank; drive two screws down into either side of the daggerboard case; and drive one more screw through the kingplank into the deckbeam slot.

PART 22: GUARDRAIL
PART 23: CUTWATER

The guardrail and cutwater are both trim pieces that provide a landing for the deck and protection for the bow of the boat.

Ingredients

- 1 × 4 × 12-foot fir or mahogany, two boards (guardrails)
- 1 × 2 clear fir, about 18 inches long (cutwater)
- Bronze or stainless steel screws (3/4-inch × #8)
- Carpenter's glue
- Marine adhesive sealant (no silicones)

Directions

1) Cut out two guardrails (each about 12 feet long) and one cutwater (about 18 inches long) from the patterns (Figures 5-22 and 5-23). As you can see from the patterns, these pieces both have two lengthwise bevels. If you lack adequate shop equipment or the experience to cut these bevels, consider having them done at a local cabinet shop. If you have access to a table saw, it's easy to set all these bevels (using a bevel square or a cardboard template); a saber saw with beveling action can also do the trick.

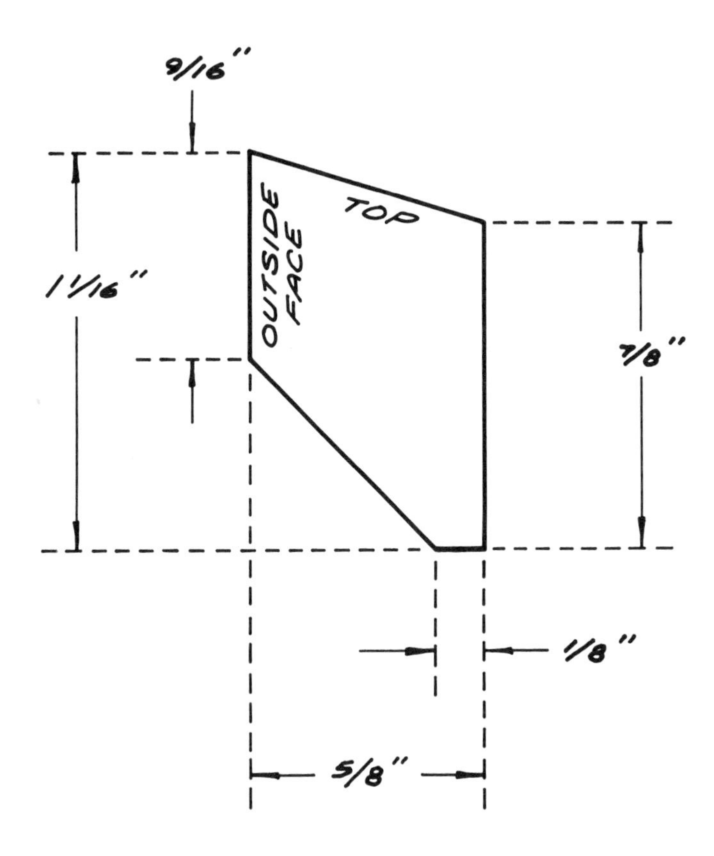

Figure 5-22. The guardrail is a trim piece that provides a landing for the boat.

2) Use clamps to position each guardrail on the outboard side of the hull, flush with the top edge. The 7/8-inch surface of the guardrail should be against the hull. Trim for length, remove the guardrail, and run a bead of glue along its surface of contact.

3) Clamp each glued guardrail back in position and secure with bronze or stainless steel screws driven every 4 inches from the inboard side through the hull out into the guardrail.

4) As for the cutwater: Position it on the stem, trim it to the right length, and modify it with plane and sandpaper so that it caps the stem and ties into the sides neatly.

5) Spread adhesive sealant on the inside face of the cutwater. Then position the cutwater and fasten it down with five bronze or stainless steel screws driven from the outside (and countersunk so that the heads can be puttied over).

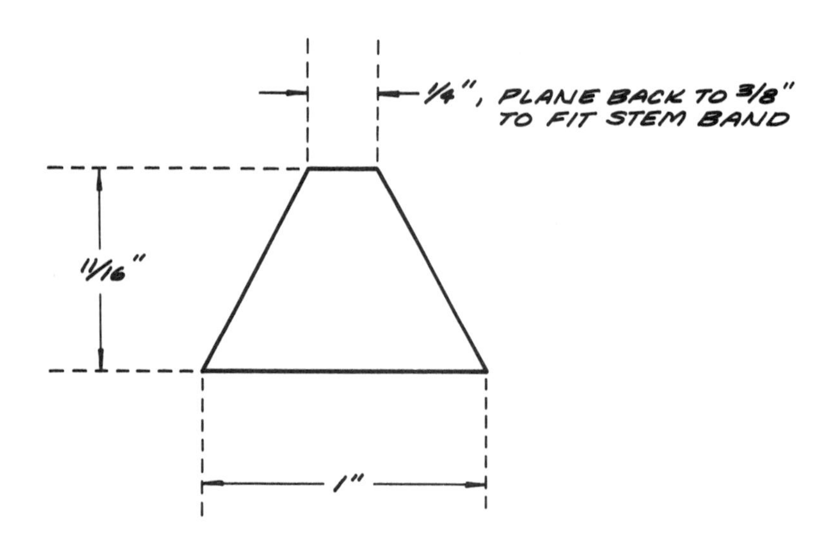

Figure 5-23. The cutwater is a trim piece that protects the bow.

PART 24: MAST STEP

The last below-deck piece is the mast step, a block that holds the mast tube. The mast slips through the deck into this tube, where it is held securely when you sail.

Ingredients

- Two $3^{1}/_{2}$ × 6-inch mahogany marine plywood blocks, $^{3}/_{4}$ inch thick
- 2-inch-diameter PVC pipe (schedule 40), 2 feet long
- Bronze screws ($2^{1}/_{2}$-inch × #12)
- Carpenter's glue
- Marine adhesive sealant (no silicones)

Directions

1) Cut out the two blocks for the mast step (from the pattern in Figure 5-24). Use a saber saw to cut a $2^{3}/_{8}$-inch-diameter hole in the upper mast-step block.

2) Glue the two mast-step blocks together, clamp tightly, and let cure overnight.

3) Next day, center the mast step on the keelson in the bottom of the boat. The mast tube should touch the forward side of the deck beam when in position. Slip the PVC pipe (mast tube) through the hole in the kingplank and down into the mast step. Butt the big square against the mast tube on top of the kingplank. The mast tube must stand above the kingplank at a 90-degree angle. Check with the square from back to front and side to side. There will be a slight rake aftward in the mast tube. Mark the forward edge of the mast step on the keelson below with a pencil. Correct positioning is illustrated in Figure 5-25.

4) Remove the tube and mast step. Sand the keelson where the mast step stood. Drill four pilot holes in the mast step, two forward and two aft of the mast hole.

5) Spread marine adhesive sealant on the bottom of the

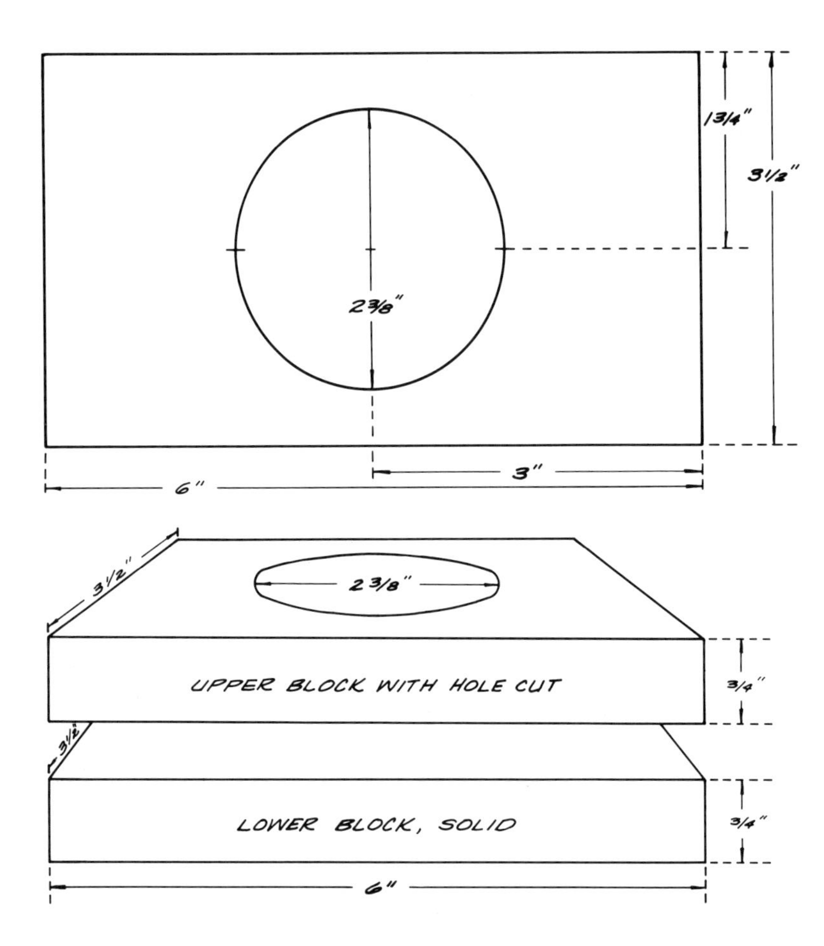

Figure 5-24. The mast step is a block that secures the mast tube below deck.

mast step, embed it in position on the keelson, and fasten it down with four bronze screws, as in Figure 5-26.

6) Spread more adhesive around the inside of the hole in the kingplank and the hole in the mast step. Slip the PVC tube down through both holes, seating it firmly.

7) Scrape away the excess adhesive and saw off the mast tube so that it is 1 inch above the kingplank.

Figure 5-25. Position the mast step to meet the mast tube and mark.

Figure 5-26. Fasten the mast step to the keelson with adhesive sealant and four screws.

Now seal the entire interior, every surface, with three (count 'em: three) coats of marine varnish or epoxy.

The interior of your Clancy is now ready to be capped with a deck.

SIX

Finishing Touches: From Deck to Daggerboard

Clancy's wraparound deck not only creates abundant flotation, it provides ideal sitting area for sailors and passengers. Two strips of deck trim are added before you paint or varnish the finished product.

After you put the deck on, we will show you how to assemble some auxiliary pieces that have to do with handling and sailing: the rudder, the tiller, and the daggerboard.

By the end of this chapter, all Clancy will lack is its mast, sail, and rigging. The basic 29 pieces will all be in place (Figure 6-1).

TOOLBOX

- Acetone
- Clamps (spring clamps, C-clamps)
- Drill (reversible) with countersink and stop collar
- Drill bits
- Handsaw
- Paint brush
- Pencil
- Plane
- Putty knife
- Saber saw and metal-cutting blade
- Scissors
- Scraper
- Speedblock sander
- Squeegee
- Tape measure
- Wood rasp

PART 25: DECK
PART 26: DECK TRIM

The deck covers all those support pieces you have been putting in, caps the flotation chambers forward and aft of the bulkheads, and provides plenty of room for comfortable seating while sailing.

Ingredients

- 1/4-inch 5-ply mahogany or fir marine plywood (two deck panels and two trim pieces)
- Bronze screws (3/4-inch × #6; 3/4-inch × #8)
- Marine adhesive sealant (no silicones)
- Epoxy resin and hardener
- Wood flour

Directions

1) Cut out two identical deck panels from the pattern in Figure 6-2.

2) Lay these panels on Clancy (Figure 6-3). The top edge of the transom must be beveled to accept the deck. Leave at least a $^1/_4$-inch gap between panels. (This gap will be covered later by the deck trim, so don't worry if it is wider than $^1/_4$ inch.) Check for fit: The panels should hang over the guardrails slightly so they later can be trimmed and sanded for an exact fit.

3) Mark the openings of the daggerboard case and the mast tube. Cut slots for each in the deck panels.

4) Paint the underside of both deck panels with epoxy.

5) Reposition the deck panels on Clancy, making sure that the slots you cut line up with the daggerboard case and mast tube. Place about 10 temporary $^3/_4$-inch × #8 screws around the deck edges and down the center.

6) Then unscrew and remove deck panels. Put down a heavy bead of marine adhesive sealant on all surfaces where the deck panels touch (Figure 6-4). Don't scrimp: Along the guardrails, for example, put down a $^1/_4$-inch-wide bead of adhesive. After the bead is down, spread it around thickly with a putty knife. You will use more than one tube of adhesive.

Figure 6-1. With decks on, Clancy is nearly ready to rig for sailing.

7) Position one panel and fasten it; then position the other and fasten it. Begin at the center of the boat and work forward and aft along the centerline, driving in and countersinking bronze screws every 6 inches.

8) After placing screws down the center of the boat,

Figure 6-2. Cut out two deck panels.

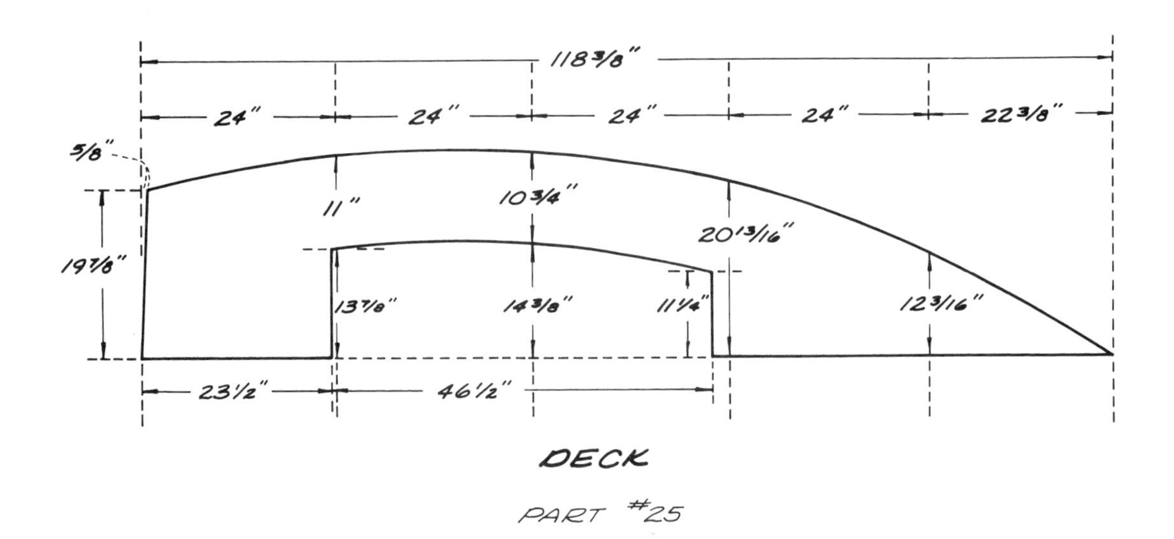

Figure 6-3. Position the deck panels on Clancy.

work crosswise over the fore and aft bulkheads, placing a screw every 3 inches.

9) Screw the deck panels down over the transom and the guardrails, driving and countersinking a screw every 6 inches (working from the center to the ends) as in Figure 6-5.

10) Finally, clean up excess adhesive with a putty knife and acetone on a rag. Then take a long breather so that the adhesive can set up.

Figure 6-4. Spread a thick bead of marine adhesive sealant over the deck support surfaces.

Figure 6-5. Screw down the deck panels every six inches on both sides of the centerline, along the carlins, across the bulkheads, and over the guardrails and transom.

When you are ready to begin again, get your fairing tools together to finish up the deck:

11) Using a saw, plane, and sandpaper, trim the deck flush with the guardrails, stem, and transom (Figure 6-6).

12) Fill all screw holes and the gap between deck panels with epoxy/wood flour putty. After the putty has set up, scrape and sand it smooth.

13) Slightly round the edges of the deck all the way around the outside of the boat with a sander.

Note: If you used fir instead of mahogany marine ply for the decking, you must sheathe it with one layer of fiberglass cloth, overlapping the guardrails and transom slightly.

Now for the final phase—trimming the deck:

14) Cut out the forward and aft deck-trim pieces (Part 26) from the patterns in Figure 6-7. For exact lengths, mea-

Figure 6-6. Trim the outboard sides of the deck panels.

sure what your own Clancy requires. The lengths given on our patterns are approximate (and generous).

15) Position the deck-trim pieces fore and aft. Then trim for fit.

16) The forward deck-trim piece must have a slot cut for the daggerboard and a hole for the mast tube (Figure 6-8).

17) Lay down a wide bead of marine adhesive sealant and embed the deck trim pieces. Deck trim can be secured by

Figure 6-7. Forward and aft deck trim.

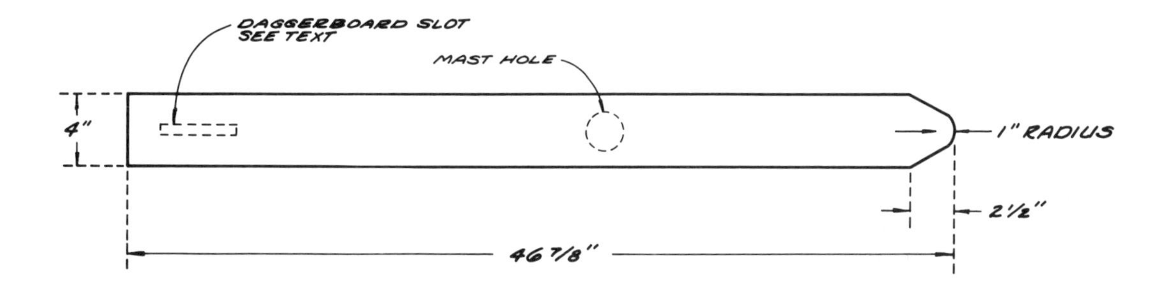

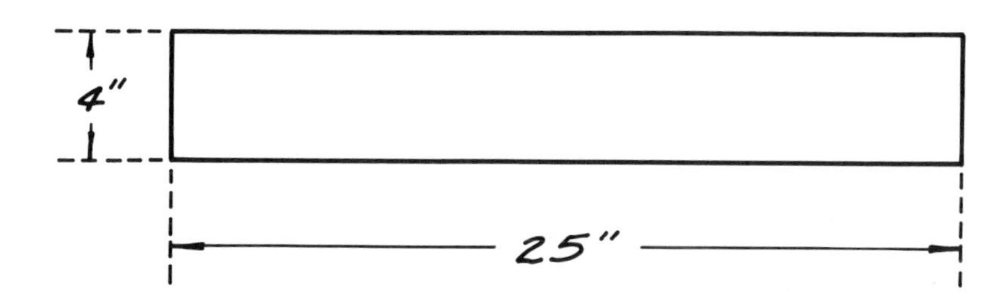

3/4-inch × #6 bronze screws (countersunk and puttied over); by brass finishing nails; or by thickened epoxy (weighted down or clamped to cure). Figures 6-9 and 6-10 show the forward and aft deck trim pieces curing in correct position.

Figure 6-8. Cut out an opening for the daggerboard and a hole for the mast tube in the forward deck-trim piece.

Figure 6-9. Clamp the aft deck trim in place to cure.

Figure 6-10. Clamp the forward deck trim in place to cure.

PAINTING CLANCY

Clancy is nearly ready for rigging. Of course, your boat still lacks a rudder, tiller, and daggerboard (not to mention mast, sail, and lines), but this is now a seaworthy vessel. So it's as good a time as any to give your Clancy the final coats of paint or varnish.

Whether you use paint or varnish is up to you. Just make sure you use real marine paints or marine varnish. Spread it on the deck, in the cockpit, along the sides, across the transom, and along the bottom – everywhere the brush could possibly reach. Follow the manufacturer's directions for proper undercoating and sealing.

Apply at least three finish coats everywhere, sanding lightly between coats.

PART 27: DAGGERBOARD AND HANDLE

It's finally time to fill that daggerboard slot through which you've had to trace and cut lo these many successive layers of boat bottom and keel, keelson and kingplank, deck and deck trim.

The daggerboard is lowered once Clancy's in the water,

of course, and provides the underwater lateral resistance for maneuverability and progress upwind when under sail.

Ingredients

- $3/4$-inch mahogany marine plywood (dagger)
- 2 × 3 × 12-inch mahogany, clear fir, or oak (handle)
- Marine adhesive sealant (no silicones)

Directions

1) Cut out the daggerboard (Figure 6-11), rounding two of the edges as shown in the pattern.

2) To help prevent drag and vibration or chattering of the daggerboard when in actual use, you can work a foil section into the lower (immersed) $2^1/2$ feet of the board with a belt or disk sander. The foil section should look like a long narrow surfboard when viewed from the bottom—$3/4$ inch thick in the middle, tapering to a sharp edge on either side. If you don't create a foil section, at least round the leading and trailing edges.

Figure 6-11. Daggerboard with view of foil area.

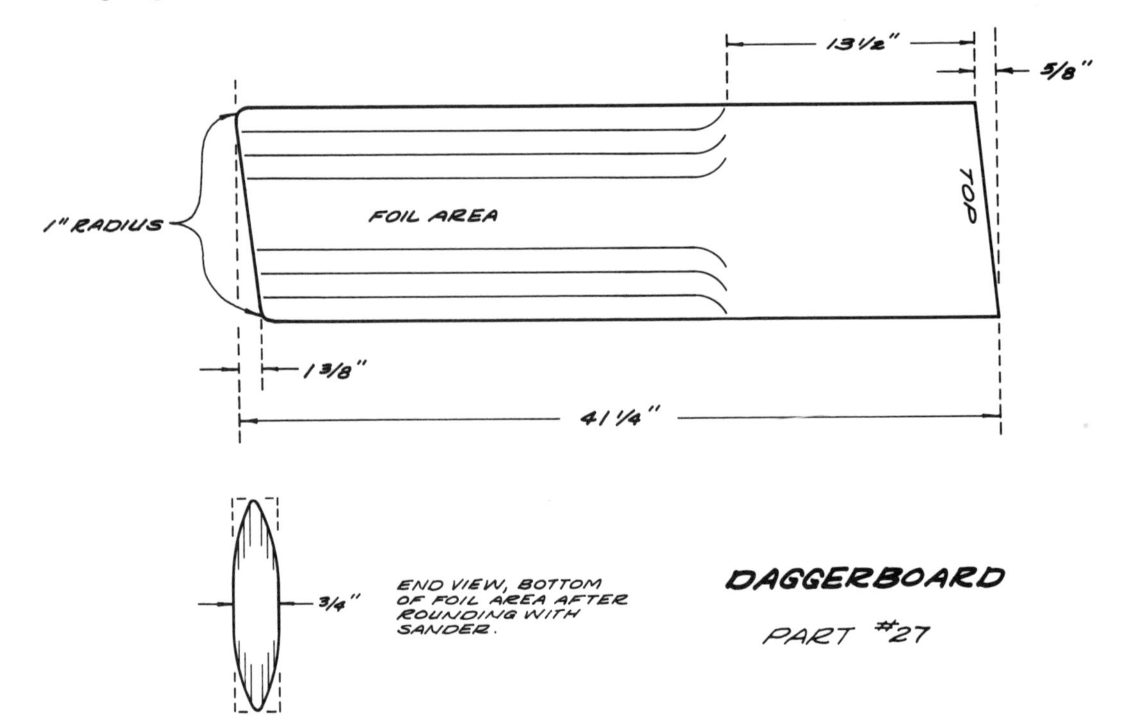

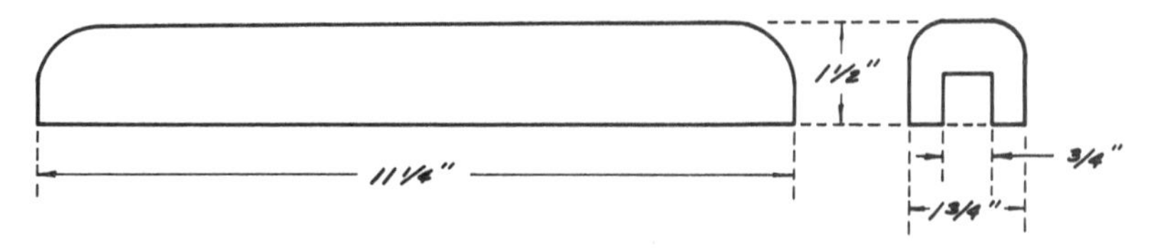

Figure 6-12. The daggerboard handle is best made of solid mahogany.

Figure 6-13. Join the handle to the daggerboard.

3) Make sure that the daggerboard drops easily all the way through the daggerboard case.

4) Fashion a handle from solid mahogany (Figure 6-12). It should fit like a tight glove over the $11^{1}/_{4}$-inch side (the side without rounded corners).

5) To glue the handle to the daggerboard, use marine adhesive sealant (Figure 6-13).

6) Sand the daggerboard and handle smooth. Paint with epoxy and at least three coats of marine varnish.

PART 28: RUDDER

The kick-up rudder enables you to come and go from shore with the rudder in place. As you come in, the rudder kicks up as it touches the sand. The rudder consists of two wooden cheeks sandwiching a marine aluminum plate that swivels on a spacer. (The rudder plate is made of marine alloy aluminum—not regular aluminum—for greater durability.)

Directions for Tracing Patterns

Figure 6-14 shows the layout of one complete rudder cheek. It is not a full-size drawing. To transfer a full-size pattern of this piece to plywood, you need to trace the three sets of lines from the next drawing (Figure 6-15) and connect them together on plywood.

The three sets of lines in Figure 6-15, when connected with each other, make up the full-size outline of the rudder cheek. The A-to-B line (which is the outline of the top half of the rudder cheek) begins at the A point in the lower left corner and runs up and around and down to the B point in the lower right corner of the pattern. The B-to-C line runs from the top to the bottom of the pattern. The C-to-A line, which runs from the lower right corner up to the top left corner of the pattern, completes the outline of the rudder cheek. Your task is to trace these three sets of lines onto wood so that they make a continuous outline of the rudder cheek, with point B in the first line connected to point B in the second line, and point C in the second line connected to point C in the third line. If you do it right, point A at the end of the third line connects to point A at the beginning of the first line.

How do you trace the pattern formed by the three sets of rudder-cheek lines in this book to your piece of plywood? One way is to photocopy or trace each set of pattern lines, cut them out as pattern pieces, tape them together (upper to lower), and trace their outline onto plywood. Another tracing method, the pin-punch approach, involves punching through the photocopied pattern lines every inch or so with a pin to score the plywood beneath.

The patterns for the rudder spacer and the rudder plate are similar to those for the rudder cheek.

The complete rudder spacer (in reduced size) is drawn in Figure 6-16. Two sets of full-size lines are contained in Figure 6-17. These two sets of lines must be joined at their A and B points to make the outline of the rudder spacer. This outline is then traced onto plywood.

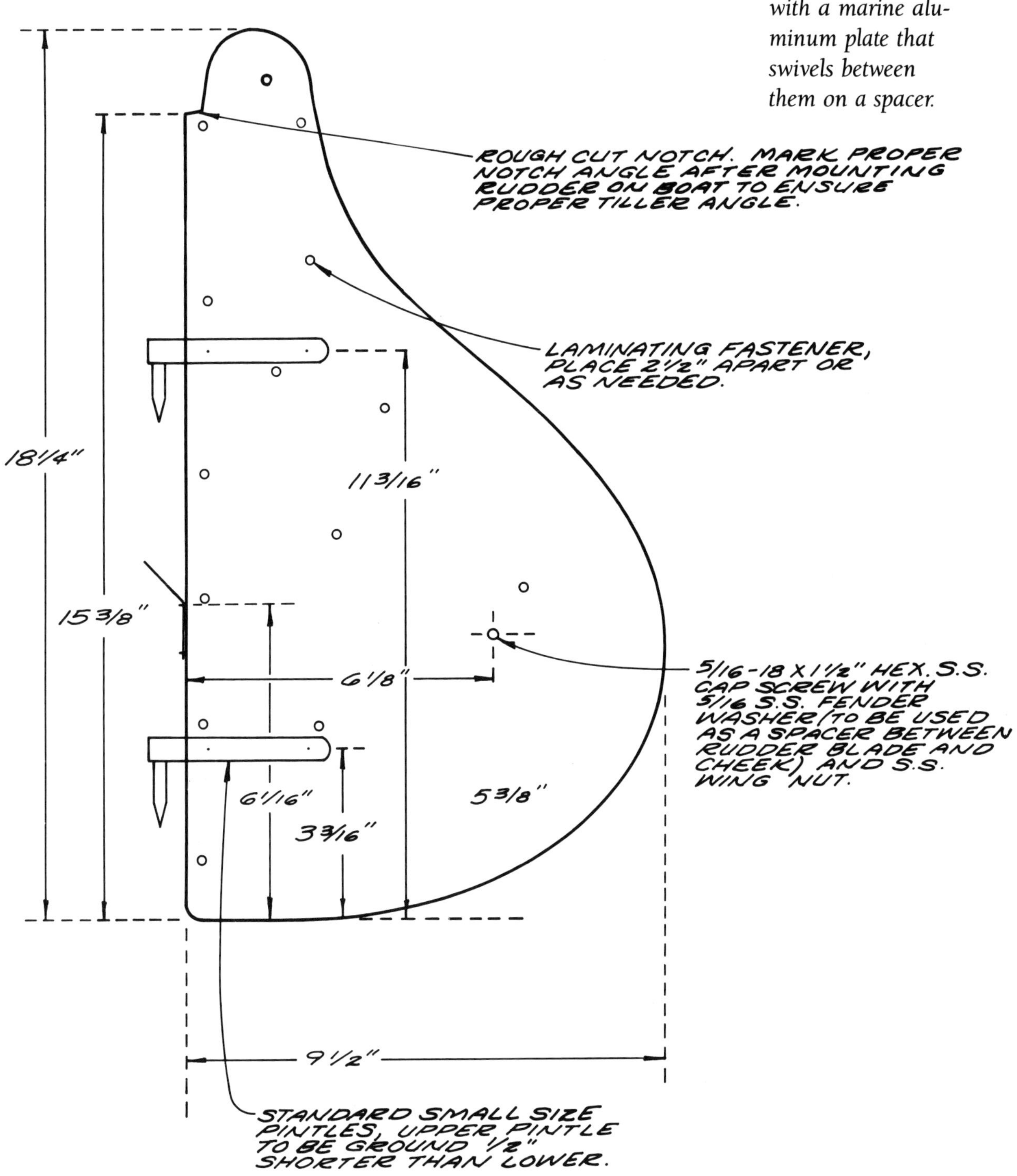

Figure 6-14. The rudder consists of two wooden cheeks with a marine aluminum plate that swivels between them on a spacer.

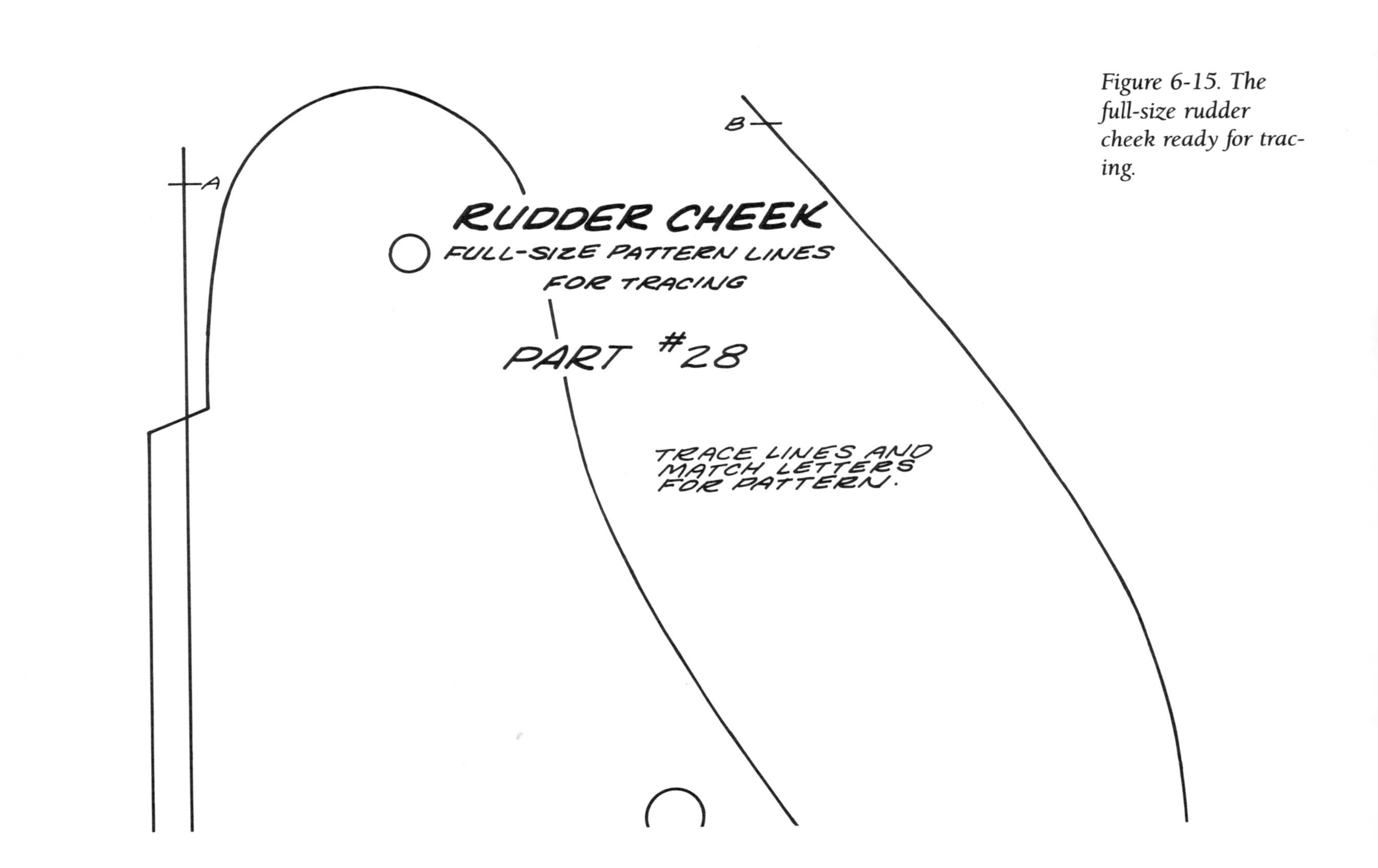

Figure 6-15. The full-size rudder cheek ready for tracing.

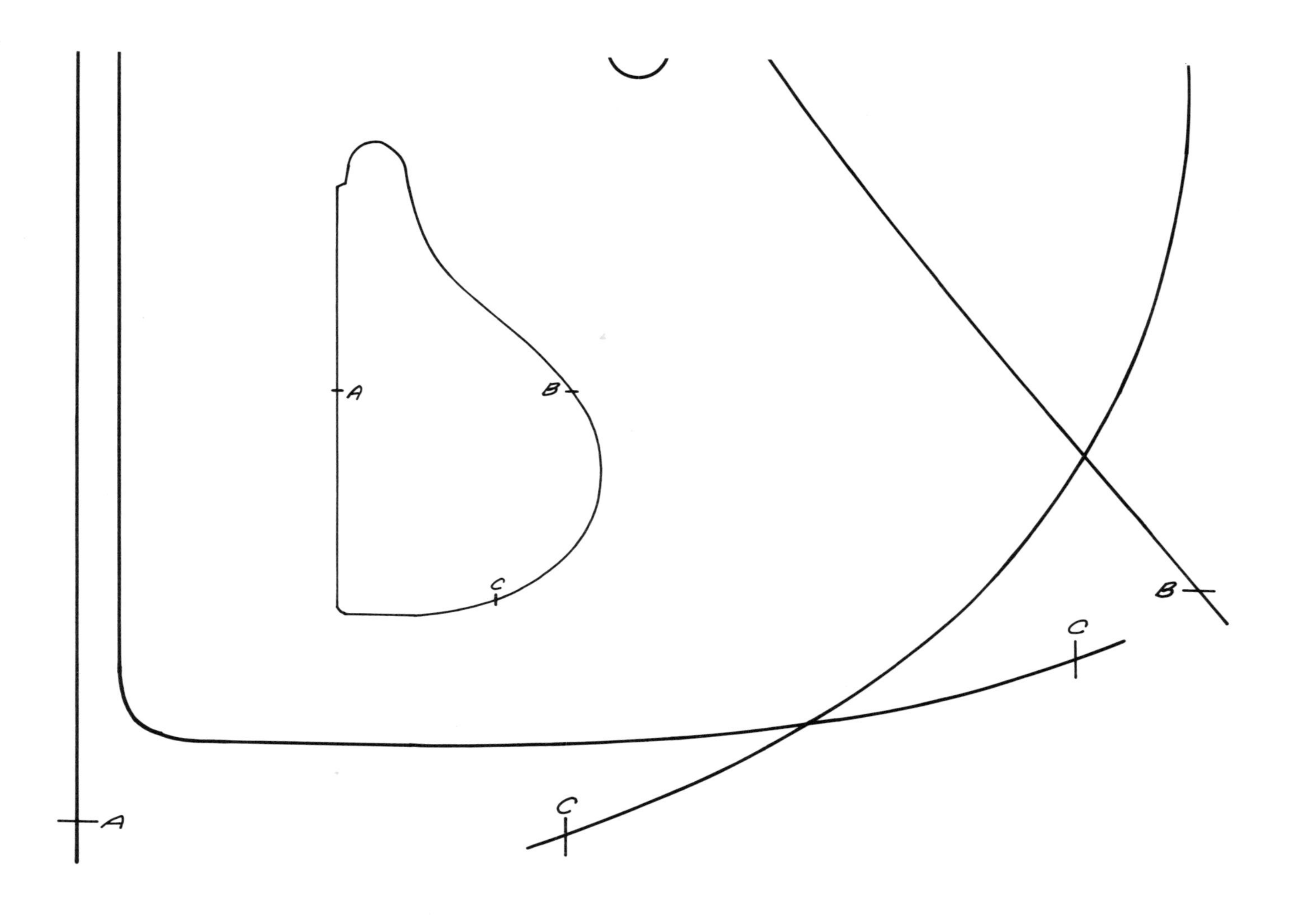
A
B
C
B
C
C
A

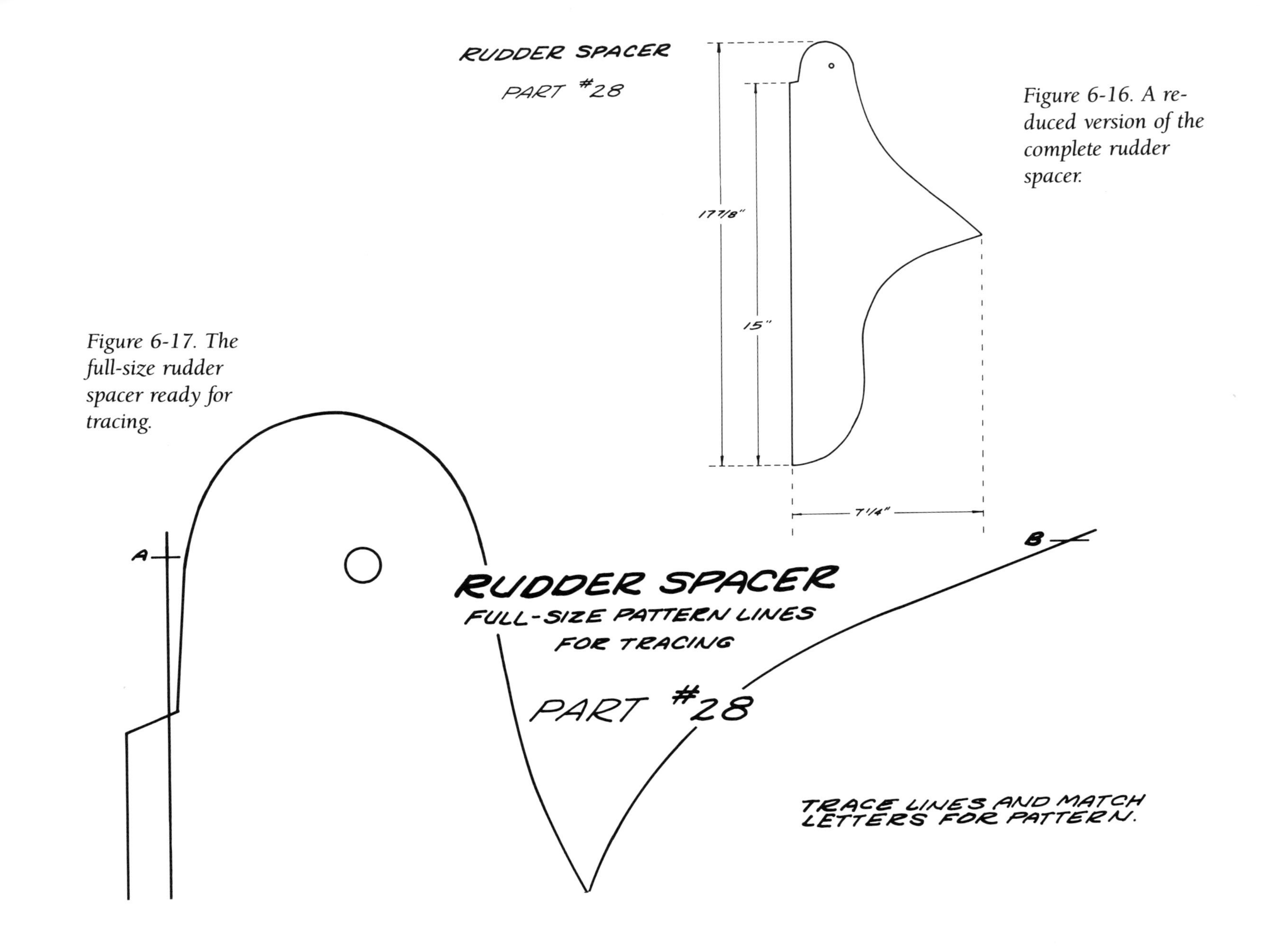

Figure 6-16. A reduced version of the complete rudder spacer.

Figure 6-17. The full-size rudder spacer ready for tracing.

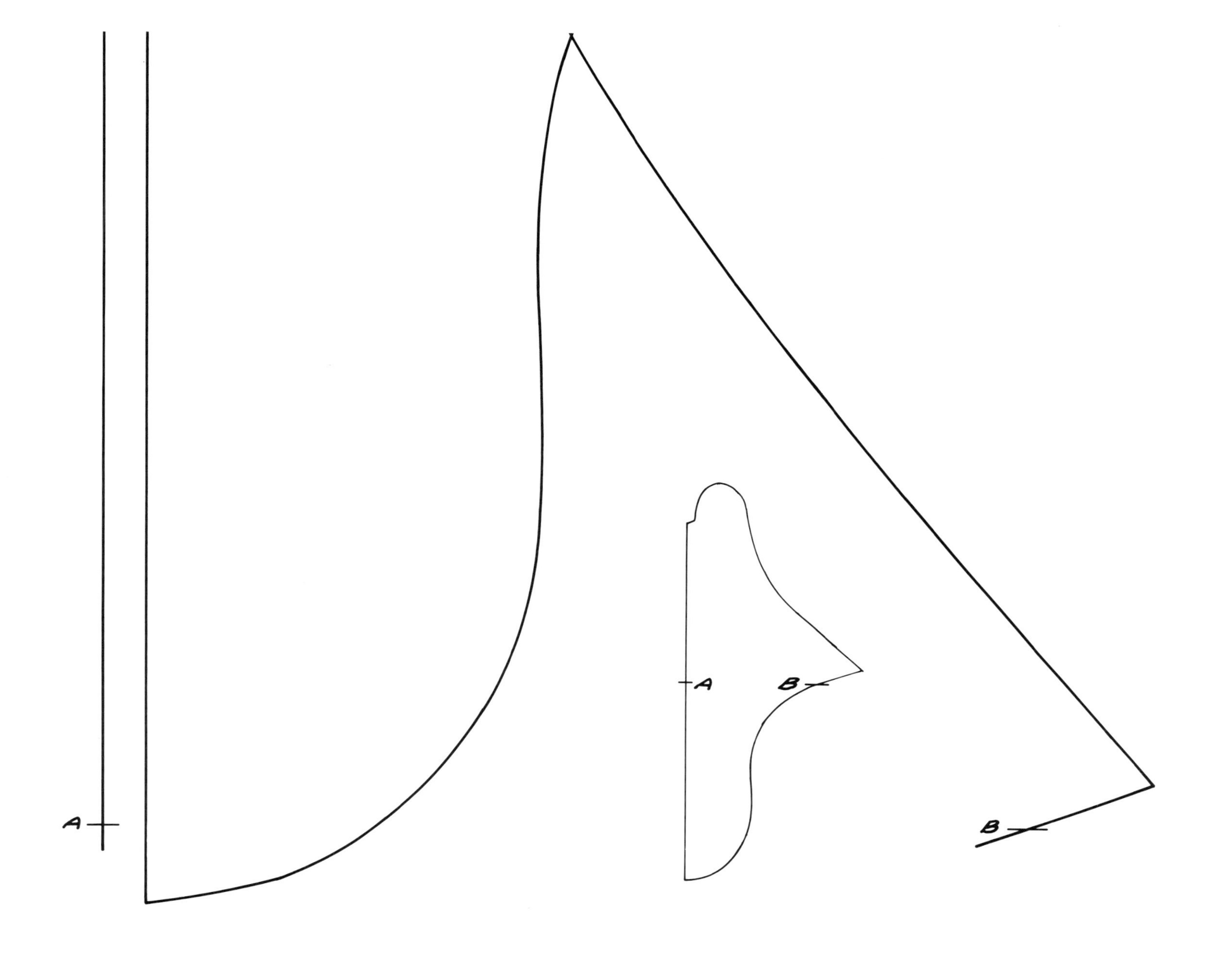

A
A
B
B

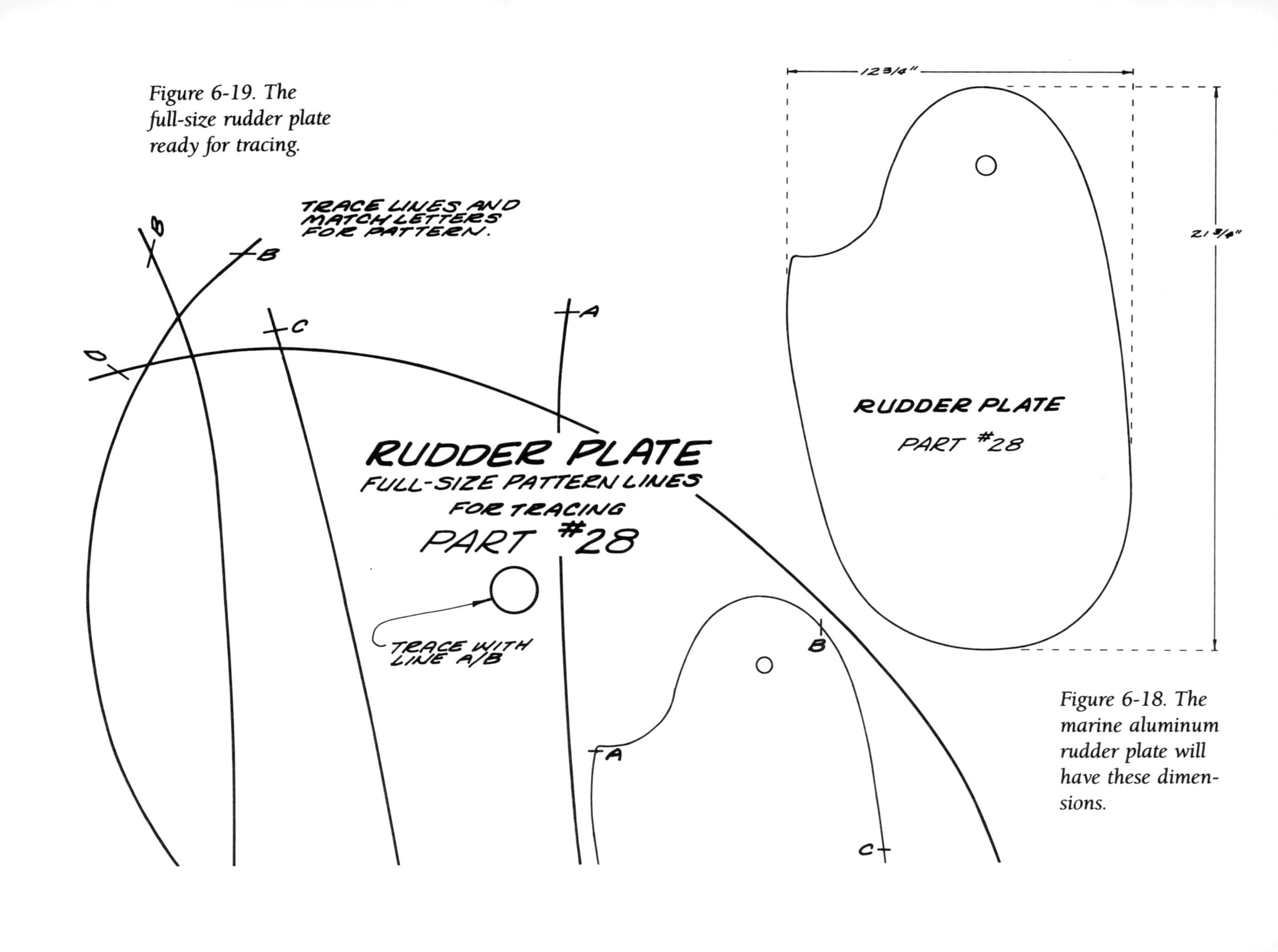

Figure 6-19. The full-size rudder plate ready for tracing.

Figure 6-18. The marine aluminum rudder plate will have these dimensions.

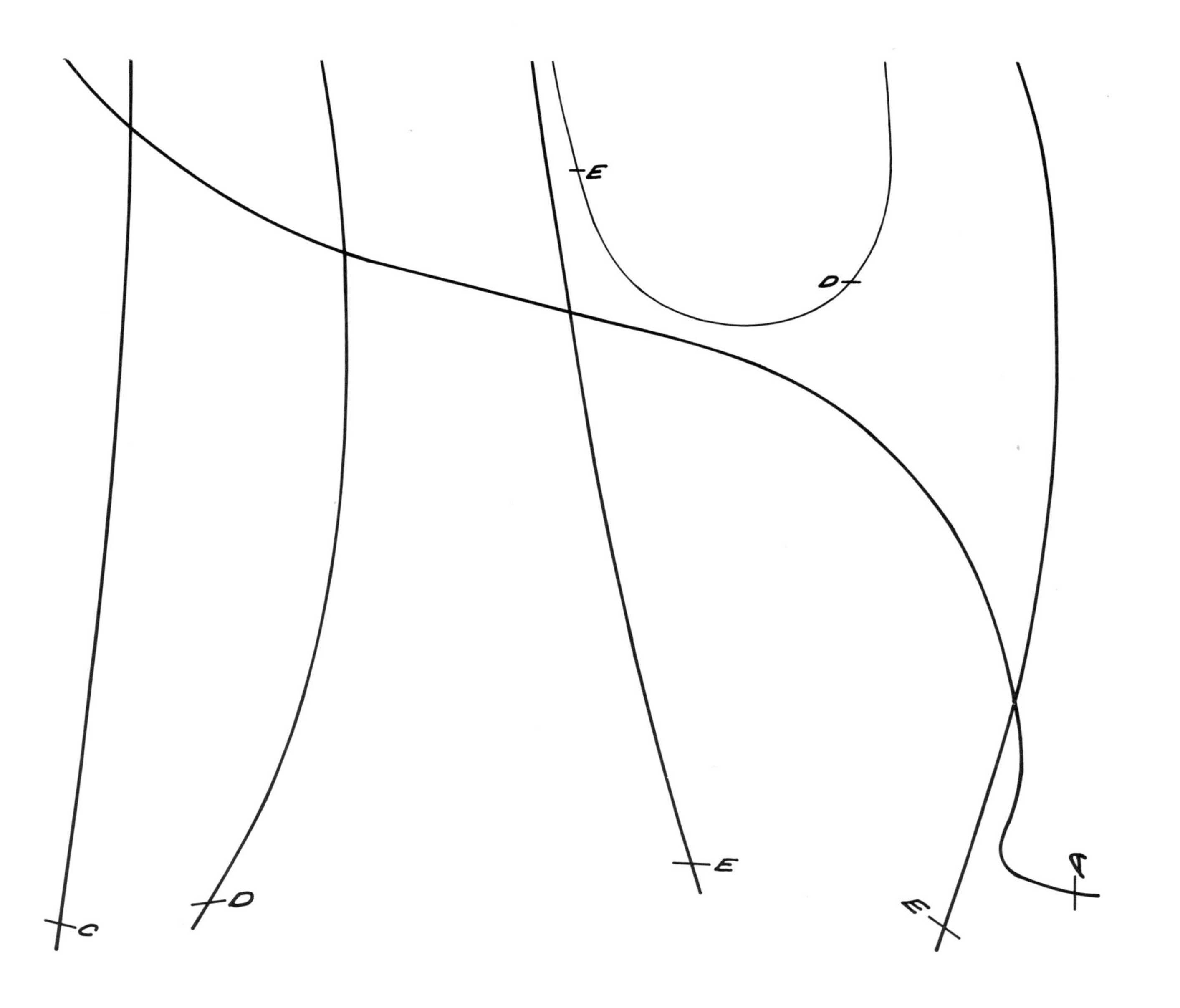

E
D
C
D
E
E
A

Finally, the rudder plate, cut from marine aluminum, is drawn (in reduced size) in Figure 6-18, with final dimensions indicated. Figure 6-19 contains the five sets of full-scale lines you need to trace, connect together, and transfer to metal.

Once the full-size lines are traced onto plywood or metal, you can cut out the pieces and assemble them into a very handy rudder.

Ingredients

- 1/4-inch 5-ply mahogany or fir marine plywood (rudder cheeks and spacer)
- 1/8-inch (4mm) marine aluminum (rudder plate)
- Two 1/16-inch-thick stainless steel 5/16-inch-diameter fender washers
- 1 stainless steel or bronze hex bolt (1 1/2-inch × #18) with same size wing nut to fit through the washers
- Bronze screws (5/8-inch × #6)
- Epoxy resin and hardener; wood flour
- Pintles and gudgeons (two standard sets)
- Roundhead stainless steel machine screws (1-inch × 10-24) with nuts and flat washers
- Roundhead stainless steel screws (1/2-inch × #10)
- 1 spring-type stainless steel rudder stop

Directions for Assembling Rudder

1) Cut out two rudder cheeks and one rudder spacer (from plywood) and one rudder plate (from marine aluminum). The parts are shown in Figure 6-20.

You can use your saber saw with a metal-cutting blade to fashion the rudder plate.

2) Lay out the cheeks so that the spacer is sandwiched between them.

3) Glue the cheeks to the spacer using an epoxy/wood flour putty. Apply this sticky mix *only where cheeks and spacer touch*. Drive 5/8-inch × #6 screws to hold cheeks together

where shown in Figure 6-14. Alternate the screws, driving the first through one cheek, the next through the other cheek, and so on.

4) Insert the metal rudder plate in the sandwich. Check that the plate pivots freely, so it can kick up when in real use.

5) Mark and drill a hole for the rudder bolt. The bolt goes through the cheeks and the metal plate, allowing the plate to pivot within the cheeks. The location of the bolt hole is indicated in the rudder patterns.

6) Push the bolt through the hole in one cheek, next through a washer, then the metal plate, then through another washer, and finally through the other cheek. Tighten the wing nut. See Figure 6-21.

7) Now test to see if the plate pivots freely with the bolt and nut in position. Adjust the tension on the bolt so that the metal plate can pivot up when Clancy is grounded, but remains down when sailing. Turning our attention now to Clancy's transom: Mount the two gudgeons on the centerline on the outside of the transom. Mount the upper gudgeon 4 inches down from the top of the transom. The lower gudgeon should be 7 inches below the upper gudgeon (11 inches from the top of the transom). See Figure 3-7. Mount the gudgeons

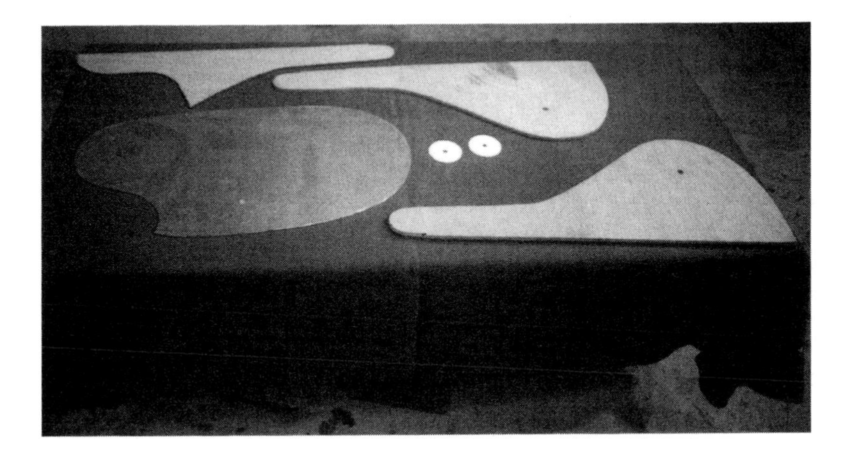

Figure 6-20. The rudder sandwich consists of two plywood cheeks, one plywood spacer, an aluminum plate, and two washers.

using 1-inch × 10-24 roundhead machine screws, secured inside the transom by nuts and flat washers.

8) Clamp the lower pintle on the rudder at the position shown in Figure 6-14. Slide the rudder into place on the transom; the lower pintle is inserted into the lower gudgeon. The upper pintle–whose pin has been ground or filed 1/2 inch shorter–is then slipped over the rudder and slid down into the upper gudgeon. Make sure that when the rudder slips into its gudgeons, it sticks up above the stern so that you can attach a tiller to the top free and clear (Figure 6-22). Mark the hole locations of the pintles on the rudder, remove the pieces, drill out the pintle holes, and use screws and adhesive to attach the pintles.

Figure 6-21. Hold the rudder sandwich together with a bolt and nut.

Figure 6-22. Position the rudder on the transom using pintle and gudgeon hardware.

9) With the rudder mounted on Clancy, mark the location of the rudder stop (above the lower pintle). The screw holes of the rudder stop should be up. Fasten the stop with 1/2-inch × #10 roundhead stainless steel screws. With the spring-type stop in place, you can remove the rudder by depressing the end of the stop.

PART 29: TILLER

The tiller connects to the rudder so that you can steer Clancy comfortably from the cockpit.

It also happens to be the final piece of your boat.

Ingredients

- 2 × 3 × 3-foot clear fir, mahogany, or oak
- 1/4-inch stainless steel or carriage bolt (2 inches long) with matching washer and elastic stop nut

Directions

1) Cut out the tiller to the dimensions shown in Figure 6-23.

Figure 6-23. The final dimensions of the tiller.

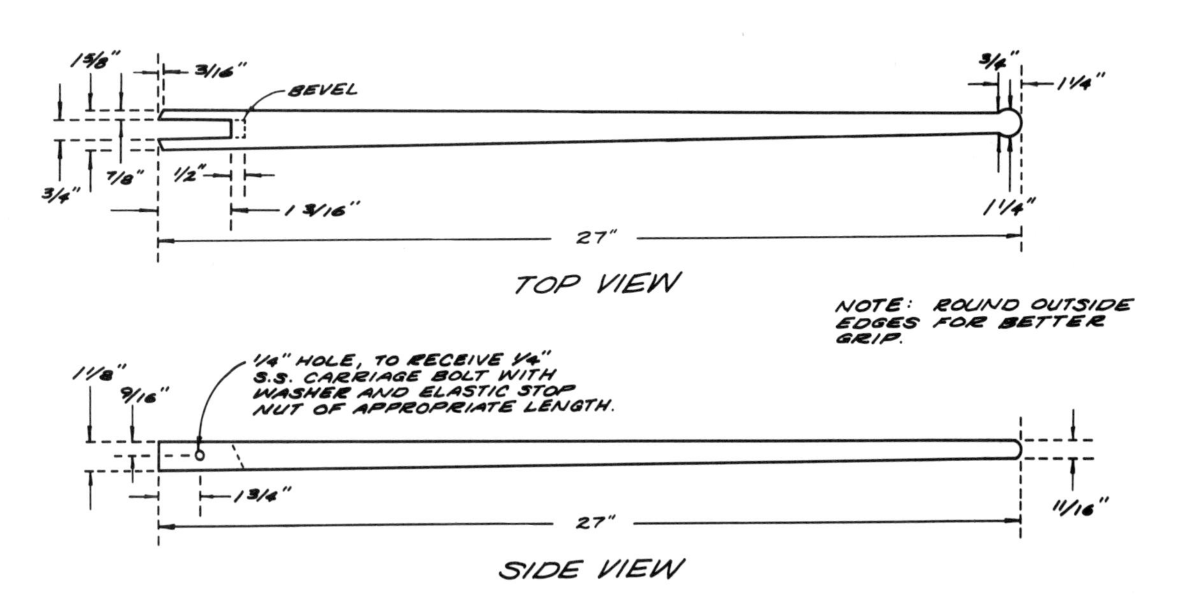

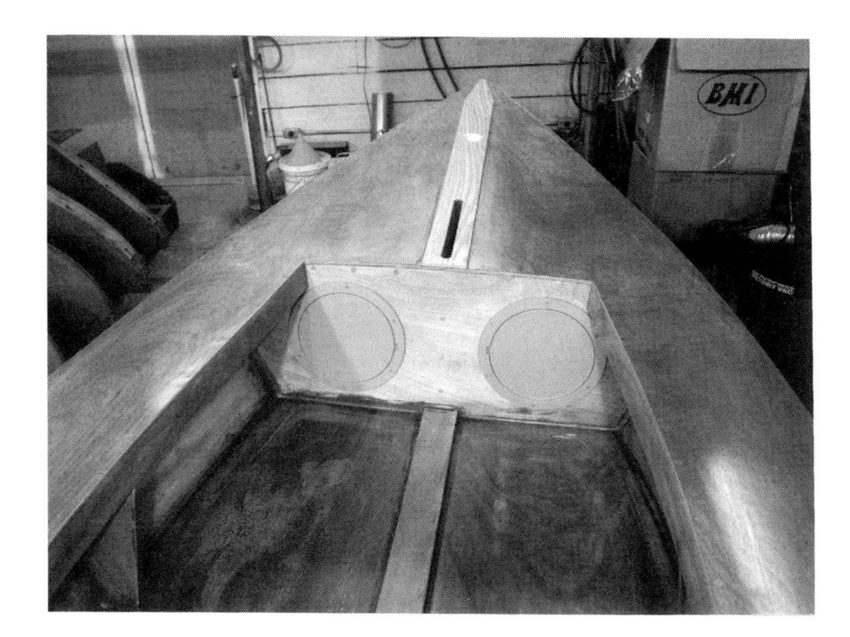

Figure 6-24. Install flanges in bulkheads.

2) Shape the tapered end of the tiller for a comfortable grip.

3) Notch the other end to fit over the rudder.

4) Drill a hole (1/4-inch) to match the hole in the top of the rudder.

5) Sand and varnish the tiller (three coats minimum).

6) Position the tiller over the rudder. Insert bolt and washer. Secure with an elastic stop nut.

INSTALL FLANGES

To create dry storage and secure the flotation chambers, you need to install a plastic flange in each bulkhead hole (two forward and one aft).

Each flange has a ring with holes around the cover. Remove the cover and try the ring in the bulkhead hole. Check for fit (the snugger, the better). Butter the opening with marine adhesive sealant and install the flange ring.

After the three rings are installed, secure the covers using $^{1}/_{2}$-inch × 10-24 stainless steel panhead bolts (Figure 6-24). Now your Clancy won't drown, and whatever you want to store forward or aft should stay dry—whether it's extra line or a picnic lunch.

With the hatches battened down, you're ready to rig your Clancy for some real sailing.

SEVEN

Rigging Clancy

Clancy is a superb sailing craft for one or two people. It is extremely stable, resists capsizing, and stays dry—perfect for the beginning sailor but bold enough to challenge the expert.

All the rigging you need to sail Clancy can be purchased off the shelf. The mast, boom, and sail are available from sailmakers, marinas, and marine supply outlets.

Because your Clancy's in a class by itself, class rules govern the mast, boom, sail, and rigging. These are covered in this chapter. In fact, our outline of how to rig Clancy not only meets the official class requirements, it constitutes the best outfit we've found for this 3-meter sailboat, whatever use you have in mind.

Whether or not you ever plan to sail Clancy in a regatta, we invite you to register with our national organization, open only to those who build Clancy from this book (just fill out the registration form in the back of the book).

TOOLBOX

- **Drill (reversible) with counter-sink and stop collar**
- **Drill bits**
- **Pencil**
- **Pop-rivet tool**
- **Screwdriver**
- **Tape measure**

THE MAST

Clancy's mast is made of 1 29/32-inch-diameter marine aluminum tubing. It is 16 feet 11 inches long and consists of two plug-together sections of equal length.

The lower mast section is 8 feet 5 1/2 inches long and fitted with an aluminum tubing insert 18 inches long. The upper section slips snugly over this insert.

The tubing insert on the inside is pop-riveted to the mast using 1/8-inch aluminum rivets. Pop-rivet tools are handy, cheap, and available at local hardware stores.

ALUMINUM
MAST AND BOOM
DECK LAYOUT AND RIGGING

Figure 7-1. Clancy Class specifications for standing and running rigging.

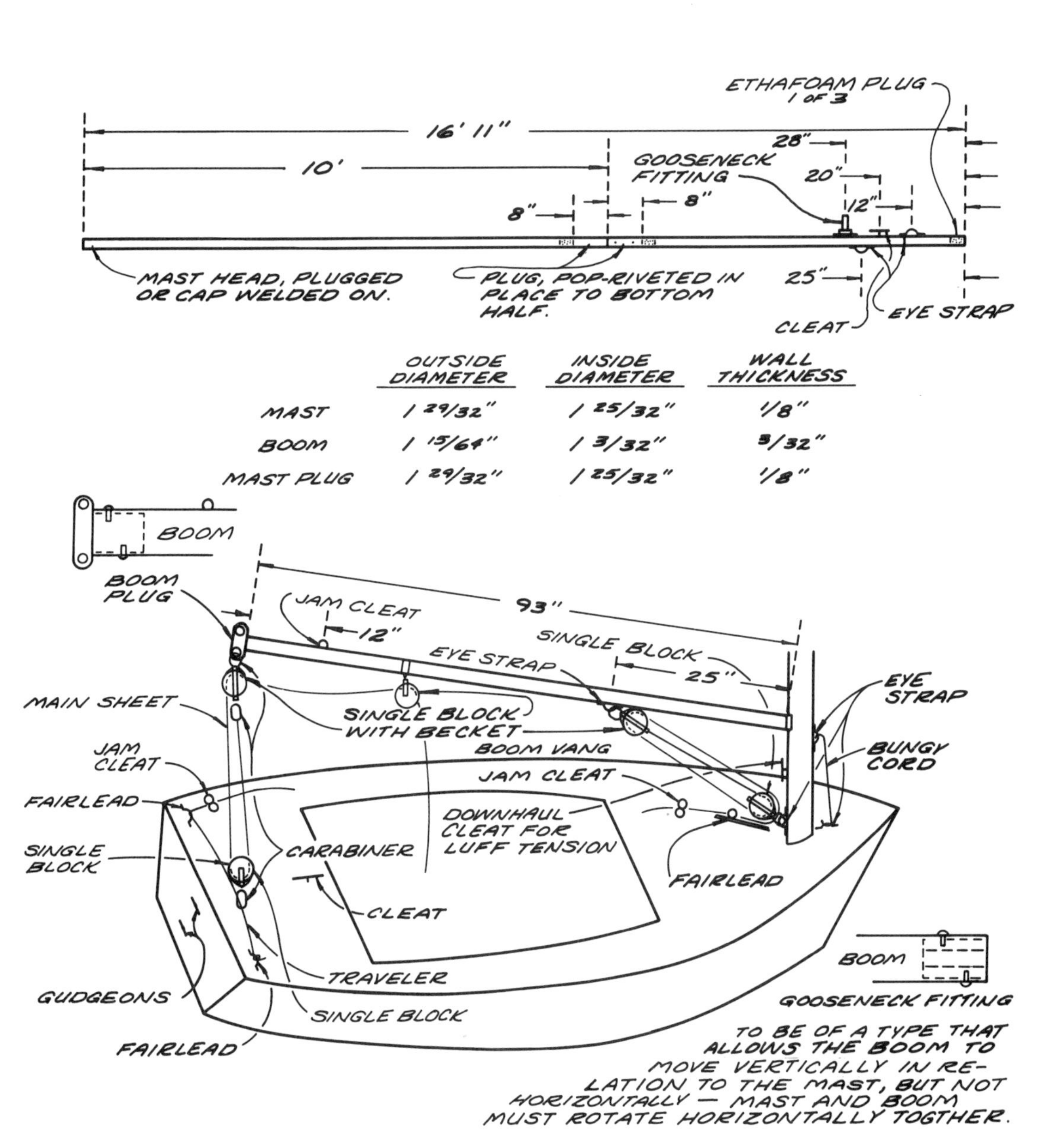

	OUTSIDE DIAMETER	INSIDE DIAMETER	WALL THICKNESS
MAST	1 29/32"	1 25/32"	1/8"
BOOM	1 15/64"	1 3/32"	3/32"
MAST PLUG	1 29/32"	1 25/32"	1/8"

The top of the mast is fitted with an aluminum cap plugged or welded on. A watertight plug of wood or plastic, glued in place, is also acceptable. Since the sail slides on over the mast, the top plug must be smooth.

The mast also contains three soft Ethafoam flotation plugs (or plugs made of any closed-cell foam) that are pushed into place inside. These provide flotation in the event that the mast inadvertently ends up in the water.

The gooseneck fitting consists of a plate-and-pin assembly. The gooseneck plate's center is riveted to the mast 28 inches from the foot.

A small 3-inch cleat is screwed to the aft face of the mast 8 inches below the gooseneck. Two small eye straps are attached to the mast, one 3 inches below the cleat and one opposite the gooseneck a few inches lower down.

All the screws used for the mast fittings are self-tapping 18-8 stainless (passive alloy). A hole slightly smaller than the screw is drilled through the mast wall, and the screw is driven in. The steel threads cut their way into the aluminum and hold tightly. Choose the shape of the screw head (flat, oval, round) by the nature of the job.

THE BOOM WITH STANDING RIGGING

The boom is made of marine aluminum tubing, 93 inches long. It's fitted with two plugs, one drilled to accept the gooseneck pin, and one cast with loops to accept the sheet block assembly and the outhaul. The plugs are held in place with two small sheetmetal screws.

A small cleat is attached 12 inches forward of the boom plug to provide tension on the outhaul.

A small snap hook is laced to the boom to accept the boom-vang block (25 inches out from the mast end of the boom).

RUNNING RIGGING

The running rigging of Clancy includes a boom traveler, the main sheet tackle, and a boom vang.

The boom traveler, located across the stern, consists of two 3/8-inch nylon fairleads and one jam cleat fastened to the aft deck. Fasten these pieces to the deck with 10-24 flathead

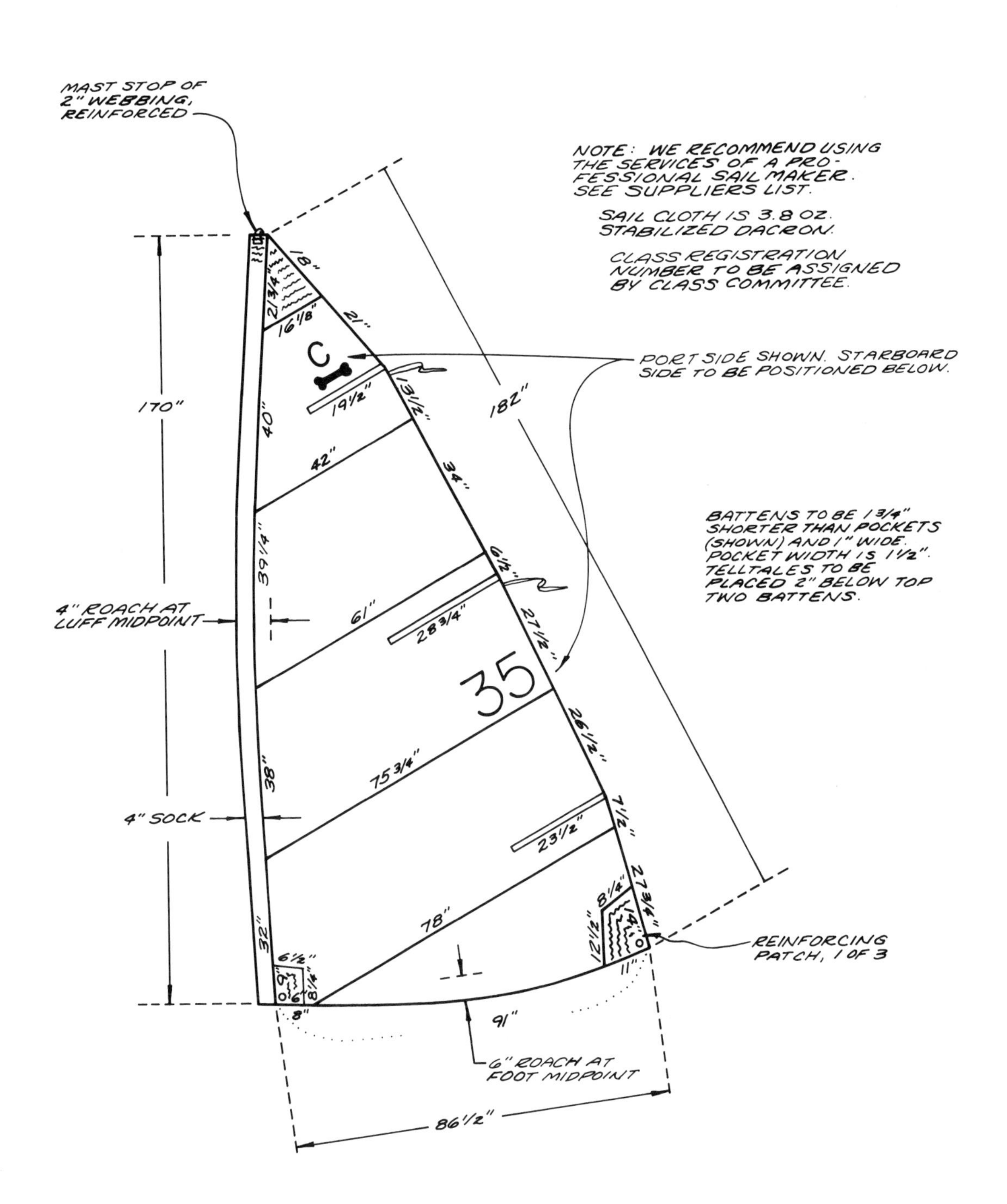

Figure 7-2. Clancy sail and C-Bone insignia.

machine screws, backed with washers and nuts under the deck. The traveler itself is a braided line 1/4 inch by 5 feet long. Tie a stopper knot at one end of this line to stop the traveler at one of the fairleads. The other end of the traveler line passes through the other fairlead into the cleat on the deck. This system allows for adjustment of the traveler's length.

The main-sheet tackle consists of the following: (1) two single swivel blocks, one with becket and one without

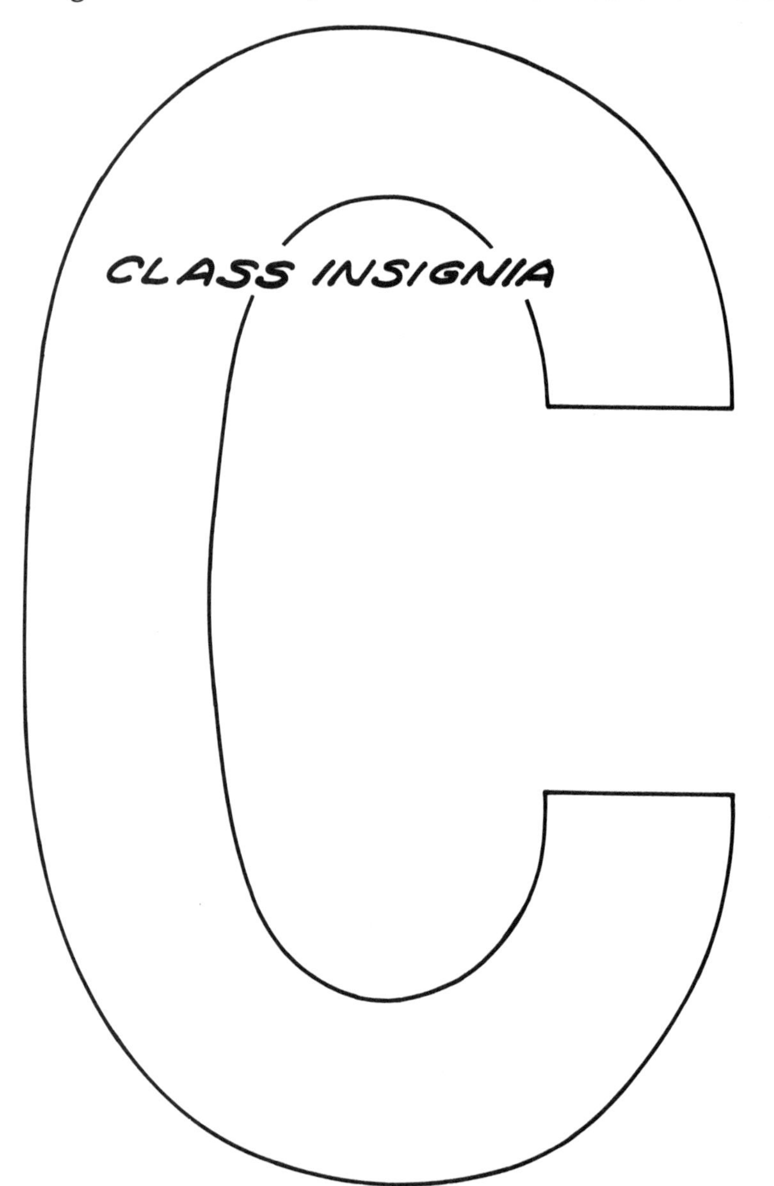

Figure 7-3. To create a full-size pattern for the C-Bone insignia, trace the "C" once and the dog bone pattern twice. Connect the two pieces at midbone.

CUT TWO. PLACE AS SHOWN ON SAIL DRAWING.

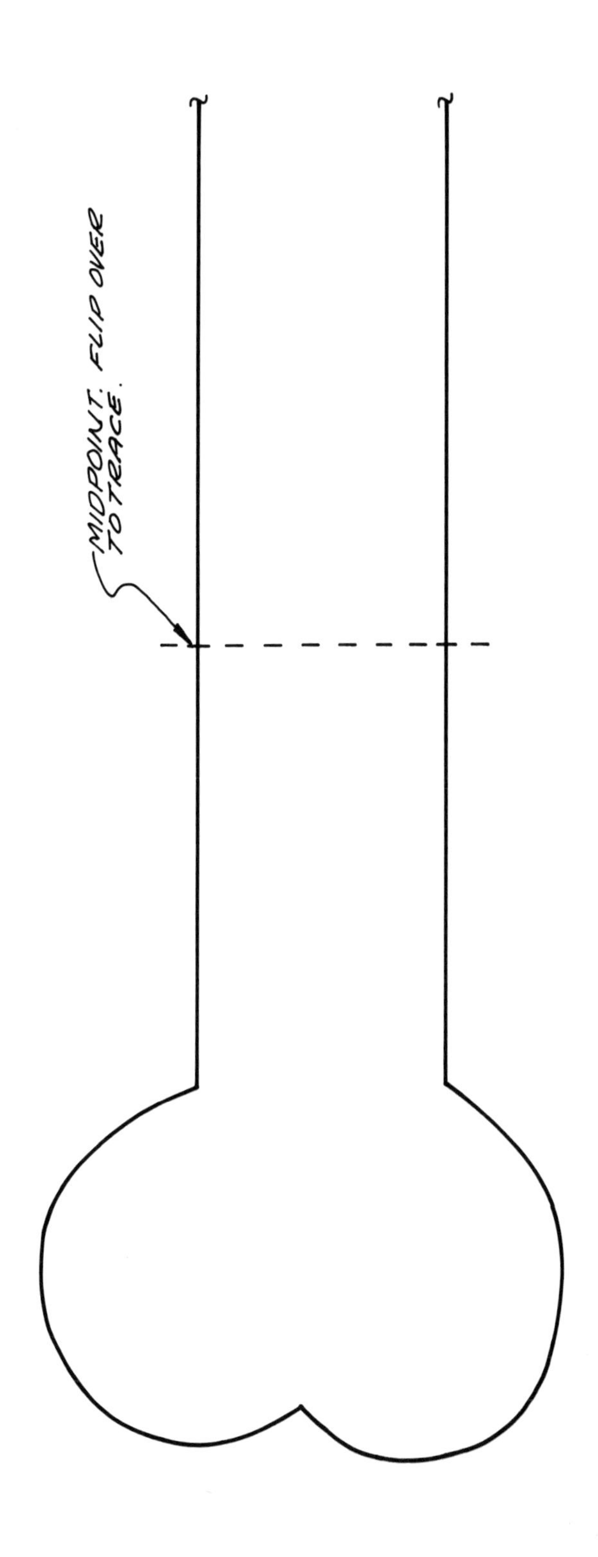
MIDPOINT: FLIP OVER TO TRACE.

becket; (2) two snap hooks; (3) one 3-inch or larger clam cleat; and (4) the main-sheet line (a 1/2-inch braided line 25 feet long with whipped ends).

The boom vang consists of one single swivel block without becket, one single swivel block with becket, one fairlead, one small cleat, and a 7-foot length of 1/4-inch braided line with whipped or heat-sealed ends.

In addition, a 1/4-inch bungy cord runs from the eye strap on the mast to the eye strap on the deck (placed 2 inches forward of the mast). This bungy cord enables the mast to rotate as you tack while sailing. It also prevents the mast from falling out of the boat in the unlikely event that Clancy takes a plunge during vigorous sailing.

The downhaul and outhaul lines are 3/16-inch braided lines, each about 2 feet long, with heat-sealed or whipped ends.

A final piece of deck hardware allows you to adjust the

(Photo by Matt Brown)

height of the daggerboard when sailing. It consists of two eye straps, a 6-inch length of 1/4-inch bungy cord, and a length of plastic tubing to cover the bungy cord. The eye straps are attached to the forward deck on either side of the daggerboard slot, each about 3 or 4 inches back from the forward edge of the daggerboard slot. Tie a stopper knot in one end of the bungy cord and feed the cord through the plastic tubing (to protect against chafing). The cord provides two functions. First, it fits over the daggerboard handle and keeps the daggerboard down in its slot. Second, it can act as a friction stop when the daggerboard is lifted up part way. Simply stretch the cord forward out of the way, lift the daggerboard part way up, and snap the cord back against the board edge to hold it in position.

THE SAIL

Clancy's sail is a loose-footed sock type, made of 3.8-ounce Dacron. The dimensions, including those for three plastic battens, are rendered on the drawing. We recommend that you use the services of a professional sailmaker.

Two colors (your choice) are customary on a Clancy sail, but the dogbone insignia is practically a requirement!

THE BONE

The class insignia, the C-Bone, is a registered trademark. You can trace our two-piece pattern onto fabric and sew two Clancy insignias, one for each side, onto your sail.

SETTING SAIL

Clancy is an extremely sporty sailboat, as you'll soon find. Be sure to register your Clancy with the national organization (there's no fee) and send us some snapshots. We can always be reached at our Clancy Hotline number or by mail at Flounder Bay Boat Lumber if you have questions or comments or need to buy materials (including a complete Clancy kit).

So let us hear from you, wherever you are. In the meantime . . .

Bone Voyage!

Glossary for Building Clancy

access ports–The holes in the bulkheads that enable you to get into the airtight compartments.

aft–Toward the back (transom end) of Clancy.

athwartships–Across the boat. A line running athwartships would cross from one side of the boat to the other.

bulkhead–A crosspiece in the boat that looks like a wall. Clancy has two of them.

carlin–A support piece running fore and aft. Clancy has two of them just under the cockpit deck.

centerline–A line running midway the length of Clancy. Technically, a plane running fore and aft that bisects a boat and is perpendicular to the waterline and the horizon.

cheeks–The wooden housing that sandwiches the metal kick-up portion of Clancy's rudder.

chine–Where Clancy's side meets Clancy's bottom.

cleat–A short stick inserted to accept a fastening or to act as a stiffener.

cockpit–The part of Clancy not decked over where you park your feet while sailing.

cockpit sole–The cockpit floor. (Clancy has considerable sole.)

crutch–A piece running fore and aft on Clancy's jig. It supports the center mold, bulkheads, and transom during construction.

cutwater–A piece that caps the stem.

daggerboard–The board that extends through Clancy to keep you from being blown sideways. A daggerboard case houses it below deck.

deck–If Clancy were a pot, the deck would be its lid.

doubler–A wooden strip on Clancy's bulkheads and partial bulkheads that reinforces or "makes double" support of the deck. (Also called a cleat.)

fastening–Nautical nomenclature for a nail, screw, or bolt.

fillet–A thick bead of putty reinforcing a corner joint. In cross section, it is concave like a cove molding.

fore–Toward the front (stem end) of Clancy.

gudgeon–Marine hardware: the socket for the rudder's pintle.

gusset–A triangular plywood reinforcement piece used to strengthen Clancy's jig.

hull–If Clancy were a pot with a lid, the hull would be the pot.

inboard–In toward the middle of Clancy from anywhere.

keel–Clancy's spine, running fore and aft across the middle of the bottom. The keel holds the two halves of the boat together.

keelson–A piece parallel to the keel and reinforcing it. In Clancy, the keelson is located inboard of the keel at your feet.

kingplank–A support plank that runs fore and aft under the center of the deck.

log–The foundation strip running fore and aft on the ladder of Clancy's jig to which other pieces are attached.

mast step–The piece below deck that holds the heel of Clancy's mast.

mast tube–In Clancy, the watertight sleeve in which the mast is stepped (located in the forward deck).

outboard–Out toward the outside of Clancy from anywhere inside Clancy.

pintle–A pin that fastens to the rudder and acts as a pivot (so that you can easily turn the rudder).

rake–An angle slightly different from 90 degrees. Clancy's stem, transom, and mast are said to be raked.

rudder–What steers Clancy from behind.

sealant–A special marine preparation used to seal wood before painting. It is not the same substance as epoxy resin.

sheer–The top edge of the boat where hull meets deck; the graceful line created thereby.

stem–Where the sides, bottom, deck, and guardrails meet in the front of the boat.

tiller–The handle connected to the rudder by which the helmsman steers Clancy.

transom–Clancy's rear end; the raked (slanted) piece where the rudder is attached.

transom knee–An angled piece that connects the keelson to the transom.

Glossary for Sailing Clancy

batten–A flat stick of wood or plastic placed in a pocket in the sail in order to keep the aft edge of the sail stiff.

boom–The pole to which the lower horizontal edge of the sail is attached. The forward end of the boom is attached to the mast.

bow–Clancy's front end (rhymes with wow).

cleat–Hardware item to which lines are "made fast" (secured).

cockpit–The part of Clancy not decked over where you park your feet while sailing.

cockpit sole–The cockpit floor. (Clancy has a lot of sole.)

daggerboard–The board that extends through Clancy to keep you from being blown sideways.

deck–If Clancy were a pot, the deck would be its lid.

downhaul–The piece of rope (always called a line on a boat) used to tighten up the forward edge of the sail (the luff) at the mast.

fairlead–A loop-like item of hardware that ensures that a line runs clear of obstacles.

gooseneck–The hardware item that connects the boom to the mast.

guardrail–The long strip of wood running along the top outboard edge of Clancy's sides. The deck is attached to the guardrail. Other common terms for the guardrail are guard, rail, outwale, and gunwale.

halyard–The rope (always called a line on a boat) used to pull (haul) or raise a sail, pennant, or boom.

hull–If Clancy were a pot with a lid, the hull would be the pot.

jam cleat–An item of hardware that holds a line fast by friction. The line is jammed in place in a jam cleat.

mast–The pole that holds the sail. It is inserted in the mast tube in the forward deck.

outhaul–The line used to pull the foot of the sail tight to the outboard end of the boom.

port–As you face the bow (forward), port is to your left (as starboard is to your right).

rudder–What steers Clancy from behind.

sail–That large piece of cloth that catches the wind and powers Clancy across the water.

sheet–The line that controls the sail.

starboard–As you face the bow (forward), starboard is to your right (as port is to your left).

stem–The front of the boat (the pointy end).

stern–The back of the boat (the squarish end).

tiller–The handle connected to the rudder by which you steer Clancy.

transom–Clancy's rear end, the raked (slanted) piece where the rudder is attached.

traveler–A track made of wood, metal, or rope on which the mainsheet pulley tackle controlling the sail "travels."

vang–A piece of tackle that prevents the boom from lifting up as the sail fills, particularly on a downwind run.

waterline–Where the water touches the hull with a "standard" 200-pound load on board.

Hardware for Clancy

Note: The stock numbers shown are from the Schaefer and SeaDog catalogs; however, Harken and Nicro-Fico make equivalent products. Any marine store that stocks sailing hardware will be able to supply what you'll need, or Flounder Bay can supply a complete hardware package.

Item	Stock number	Fasteners
1 black aluminum cleat (5 inches)	SCH #70-95	use 2, 1-inch × #12 FH SS MS
1 black aluminum cleat (3 inches)	SCH #70-13	use 2, 1-inch × #8 FH SS ST
5 eye straps	SCH #78-32	use 10, $^{1}/_{2}$-inch × #10 RH SS ST
3 fairleads	SCH #78-51	use 6, 1$^{1}/_{2}$-inch × #10 RH SS MS, nuts, and washers
3 clam cleats	SDL #002090	use 2, 1$^{1}/_{2}$-inch × #8 FH SS ST for boom mount 4, 1$^{1}/_{2}$-inch × #8 FH SS MS, nuts, and washers
3 carabiners or snap hooks	SDL #151060	

Blocks for sheet ($^{1}/_{2}$-inch line)	
2 SCH #03-05	Series 3 single block with swivel
1 SCH #03-15	Series 3 single block with swivel and becket
Blocks for boom vang	
1 SCH #01-17	Series 1 single block with swivel and becket
1 SCH #01-07	Series 1 single block with swivel
Lines for running rigging	
Mainsheet	25 feet of $^{1}/_{2}$-inch diameter nylon or Dacron braided line with whipped ends
Traveller	5 feet of $^{1}/_{4}$-inch braided nylon or Dacron with whipped ends
Downhaul and outhaul	$^{3}/_{16}$-inch braided nylon or Dacron, about 2 feet each with heat-sealed or whipped ends
Boom vang	7 feet of $^{1}/_{4}$-inch braided nylon or Dacron, whipped or heat-sealed; snap shackles may be permanently attached to mainsheet and vang for convenience

Rudder fittings

2 light-duty gudgeons

SCH #81-01	use 6, 3/4-inch × #10 RH SS MS, nuts, and washers; for bottom gudgeon 2, 1 1/2-inch × #10 RH SS WS may be needed if transom knee is 1 1/2 inches thick or thicker

2 light-duty pintles for use with a 3/4-inch rudder

SCH #80-11	1 1/2-inch pin length (bottom)
SCH #80-12	1 1/4-inch pin length (top)
	use 4, 1-inch × #10 RH SS MS, nuts, and washers

1 spring-type SS rudder stop

SCH #82-01	use 2, 1/2-inch × #10 RH SS WS

(Mark the location of the stop with rudder mounted on the boat. The stop should be located above the lower pintle, with the mounting screws on top. To remove the rudder, depress the end of the stop.)

Stem band

3/8-inch half-oval brass; 12-foot length; secure with 5/8-inch × #5 brass OH WS on 6-inch centers; 24 needed

Guards

1/2-inch half-oval brass; 2, 12-foot lengths; fasten on 6-inch centers with 5/8-inch × #5 brass OH WS; 48 needed

Abbreviation key:

FH	flat head	SS	stainless steel
RH	round head	ST	self-tapping
OH	oval head	WS	wood screws
MS	machine screws		

Suppliers

All the materials to build Clancy (lumber, hardware, rigging, and sails) are available individually or as a package from:

Flounder Bay Boat Lumber
1019 Third Street
Anacortes, WA 98221
(206) 293-2369 or (800) 228-4691

Miscellaneous wood is available from Flounder Bay or your local lumberyard. Paints, epoxies, marine adhesive sealants, and fiberglass products are available from Flounder Bay or your local marine dealer.

Plywood is also available from:

Harbor Sales
1401 Russell Street
Baltimore, MD 21230
(800) 345-1712

M.L. Condon Co.
260 Ferris Avenue
White Plains, New York 10603
(914) 946-4111

Mast and boom sets (the Clancy set) are also available from:

Yacht Riggers
4448 27th West
Seattle, WA 98199
(206) 282-7737

Hardware is also available from:

The Crow's Nest
1900 N. Northlake Way
Seattle, WA 98103
(206) 632-3555

Doc Freeman's, Inc.
999 N. Northlake Way
Seattle, WA 98103
(206) 633-1500

Hamilton Marine
155 East Main Street
Searsport, Maine 04974
(207) 548-6302

Sails are also available through:

Shore Sails
1607 Dexter Avenue North
Seattle, Washington 98109
(206) 284-3730

Registration numbers are available through:

Association Headquarters
1019 Third Street
Anacortes, WA 98221
(206) 293-2369

Clancy Hotline

(206) 293-2369
FAX: (206) 293-4749

Open 8:30 a.m. - 5:30 p.m., Pacific Time
Monday through Saturday

By Mail:

Clancy
Flounder Bay Boat Lumber
1019 Third Street
Anacortes, WA 98221
U.S.A.

Index

Clancy Class Official Registration Form

Upon completion of your Clancy, please fill in this application form to enroll your boat in the national organization. There is no membership fee. We record the data you provide on our permanent register and assign your boat an official builder's number.

- -

Clancy Builder's Official Registration Form and Class Verification

BUILDER'S NAME ____

ADDRESS ____

____

PHONE NUMBER ____

BOAT'S NAME ____

DATE BEGUN/DATE COMPLETED ____________________ / ____________________

DATE/PLACE OF INAUGURAL LAUNCH ____________________ / ____________________

PLYWOOD USED: MAHOGANY __________ FIR __________ OTHER __________

COMMENTS, CRITICISMS, COMPLIMENTS, ETC. ______________________________

____

____

Send to: Clancy
Flounder Bay Boat Lumber
1019 Third Street
Anacortes, WA 98221